MONOGRAPHS ON
STATISTICS AND APPLIED PROBABILITY

General Editors

D.R. Cox, D.V. Hinkley, D. Rubin and B.W. Silverman

The Statistical Analysis of Compositional Data
J. Aitchison

Probability, Statistics and Time
M.S. Bartlett

The Statistical Analysis of Spatial Pattern
M.S. Bartlett

Stochastic Population Models in Ecology and Epidemiology
M.S. Bartlett

Risk Theory
R.E. Beard, T. Pentikäinen and E. Pesonen

Bandit Problems Sequential Allocation of Experiments
D.A. Berry and B. Fristedt

Residuals and Influence in Regression
R.D. Cook and S. Weisberg

Point Processes
D.R. Cox and V. Isham

Analysis of Binary Data
D.R. Cox

The Statistical Analysis of Series of Events
D.R. Cox and P.A.W. Lewis

Analysis of Survival Data
D.R. Cox and D. Oakes

Queues
D.R. Cox and W.L. Smith

Stochastic Modelling and Control
M.H.A. Davis and R. Vinter

Stochastic Abundance Models
S. Engen

The Analysis of Contingency Tables
B.S. Everitt

An Introduction to Latent Variable Models
B.S. Everitt

Finite Mixture Distributions
B.S. Everitt and D.J. Hand

Population Genetics
W.J. Ewens

Classification
A.D. Gordon

Monte Carlo Methods
J.M. Hammersley and D.C. Handscomb

Identification of Outliers
D.M. Hawkins

Generalized Linear Models
P. McCullagh and J.A. Nelder

Distribution-free Statistical Methods
J.S. Maritz

Multivariate Analysis in Behavioural Research
A.E. Maxwell

Applications of Queueing Theory
G.F. Newell

Some Basic Theory for Statistical Inference
E.J.G. Pitman

Density Estimation for Statistics and Data Analysis
B.W. Silverman

Statistical Inference
S.D. Silvey

Models in Regression and Related Topics
P. Sprent

Regression Analysis with Applications
G.B. Wetherill

Sequential Methods in Statistics
G.B. Wetherill and K.D. Glazebrook

(Full details concerning this series are available from the Publishers)

Asymptotic Techniques

for Use in Statistics

O.E. BARNDORFF-NIELSEN

Department of Theoretical Statistics
Aarhus University, Aarhus, Denmark

and

D.R. COX

Nuffield College, Oxford, UK

LONDON NEW YORK
CHAPMAN AND HALL

First published in 1989 by
Chapman and Hall Ltd
11 New Fetter Lane, London EC4P 4EE
Published in the USA by
Chapman and Hall
29 West 35th Street, New York NY 10001

Set in 10/12 pt Times by Thomson Press (India) Ltd, New Delhi
Printed in Great Britain by J.W. Arrowsmith, Bristol

ISBN 0 412 31400 2

British Library Cataloguing in Publication Data

Barndorff-Nielsen, O.E. (Ole Eiler), *1935–*
Asymptotic techniques.
1. Asymptotic statistical mathematics
I. Title II. Cox, D.R. (David Roxbee), *1924–* III. Series
519.5

ISBN 0 412 31400 2

Library of Congress Cataloging in Publication Data

Barndorff-Nielsen, Ole.
Asymptotic techniques for use in statistics/O.E. Barndorff-Nielsen and D.R. Cox.
p. cm.—(Monographs on statistics and applied probability)
Bibliography: p.
Includes index.
ISBN 0 412 31400 2
1. Mathematical statistics—Asymptotic theory. I. Cox, D.R. (David Roxbee) II. Title. III. Series.
QA276.B2846 1989
519.5—dc 19

Contents

Preface

The use in statistical theory of approximate arguments based on such methods as local linearization (the delta method) and approximate normality has a long history. Such ideas play at least three roles.

First they may give simple approximate answers to distributional problems where an exact solution is known in principle but difficult to implement. The second role is to yield higher-order expansions from which the accuracy of simple approximations may be assessed and where necessary improved. Thirdly the systematic development of a theoretical approach to statistical inference that will apply to quite general families of statistical models demands an asymptotic formulation, as far as possible one that will recover 'exact' results where these are available.

The approximate arguments are developed by supposing that some defining quantity, often a sample size but more generally an amount of information, becomes large: it must be stressed that this is a technical device for generating approximations whose adequacy always needs assessing, rather than a 'physical' limiting notion.

Of the three roles outlined above, the first two are quite close to the traditional roles of asymptotic expansions in applied mathematics and much of the very extensive literature on the asymptotic expansion of integrals and of the special functions of mathematical physics is quite directly relevant, although the recasting of these methods into a probability mould is quite often enlightening.

The third role mentioned above is much more specialized and will not be covered in the present volume. Rather we have aimed to set out the main mathematical methods useful in handling asymptotic calculations in statistical theory, and to a limited extent in applied probability. A difficulty that has to be faced in writing about this topic is the clash between the description of formal methods and a careful statement of the precise regularity conditions that provide the theorems that justify such manipulations. The formal methods are often simple and intuitively appealing, whereas the same cannot be said about the regularity conditions and rigorous proofs! We have aimed to resolve this difficulty by concentrating on the formal methods and their exemplification in simple and not so simple examples, giving the appropriate theorems in appendices at the ends of chapters, with, however, at most an outline proof.

The first four chapters deal with one-dimensional problems and the last three chapters with the corresponding multidimensional techniques. In most cases it is possible, by choice of suitable notation, to show the multidimensional results as direct generalizations of the corresponding univariate ideas. Implementation in multidimensional problems can, all the same, involve heavy calculations and it is likely that the use of computerized algebra would often be helpful, although this is a topic we do not discuss.

We are very grateful to P. Hall and R. Gill for advice over some points in Chapters 4 and 6 respectively and to V.S. Isham, J. L. Jensen, P. McCullagh, N. Reid and I. M. Skovgaard for reading a preliminary version and making exceptionally constructive and detailed comments. Of course the errors that remain are solely our responsibility!

Two secretaries at the University of Aarhus, Linda Allermand and Oddbjørg Wethelund, did work of outstanding quality on a difficult manuscript.

Our collaboration has been greatly helped by travel grants from NATO and later from the EEC. We have been supported throughout by the Danish Science Research Council (O.E.B.N.) and by an SERC Senior Research Fellowship (D.R.C.), the latter held at the Department of Mathematics, Imperial College, London. We acknowledge all this support with gratitude.

Finally, we dedicate this book to Bente and Joyce for their care, support and tolerance.

O.E. Barndorff-Nielsen
Aarhus

D.R. Cox
London

June 1988

CHAPTER 1

Preliminary notions

1.1 Introduction

Some of the distributional problems arising in the theory of statistics have usable exact solutions. Other problems, including many of practical interest, either have no exact solution or have solutions that are so complicated as to be inconvenient for direct use. Such situations are usually handled by so-called asymptotic or large-sample theory. That is, we derive approximations by supposing that some quantity n, usually a sample size, but more generally a quantity of information, is large.

For example, the limit laws of probability theory may show that as $n \to \infty$ a distribution of interest approaches some simple standard distribution, in particular the normal distribution associated with the central limit theorem. More thorough analysis may show that the normal distribution is the leading term of an asymptotic series, often in powers of $n^{-1/2}$ or n^{-1}. The leading term on its own may give an adequate numerical approximation; the further terms may be used either to improve that approximation or to assess the conditions under which the simple approximation is likely to be serviceable.

Such expansions may well have rigorously established mathematical properties holding as $n \to \infty$ but, as in other areas of application, they are to be regarded here as ways of deriving approximations to be used in practice for particular finite values of n and the question of the adequacy of the approximation always needs attention.

The object of the present book is to discuss systematically the mathematical techniques of asymptotic theory useful in statistics and applied probability. So far as is feasible, we have deferred careful discussion of analytical regularity conditions to appendices. Our object is to concentrate attention on the key ideas and to make the discussion reasonably accessible to those primarily interested in applying the results. Throughout, the account is illustrated by examples, most of which are intended to be of some intrinsic interest.

In the first part of the book we deal primarily with univariate random variables, formulating results in a notation chosen as far as is feasible to ease passage to the multivariate case considered later.

Our interest in these topics arose largely, although not entirely, because of the implications for the general theory of parametric statistical inference. It is hoped to discuss that particular topic in detail in a subsequent volume.

1.2 Sums of independent random variables: standardization

It is natural to consider first sums of independent random variables, in particular because many of the key ideas are most simply illustrated in this way. Let $Y_1, Y_2, \ldots$ be independent and identically distributed random variables, copies of a random variable Y. Unless explicitly stated otherwise, we assume that moments of all orders exist and in particular we write

$$E(Y) = \mu, \qquad \text{var}(Y) = \kappa = \sigma^2. \tag{1.1}$$

Consider

$$S_n = Y_1 + \cdots + Y_n; \tag{1.2}$$

then

$$E(S_n) = \mu n, \qquad \text{var}(S_n) = \kappa n = \sigma^2 n. \tag{1.3}$$

The first step in virtually all asymptotic calculations is to standardize the random variable of interest, i.e. to apply a linear transformation (change of origin and scale) so that nondegenerate limiting distributions are obtained as $n \to \infty$. Very often this is done so that the new random variable has, exactly or approximately, zero mean and constant, or even unit, variance. Thus in the present case we may put

$$S_n^* = (S_n - n\mu)/(\sigma\sqrt{n}) \tag{1.4}$$

or may omit σ in the denominator and consider

$$S_n^* = (S_n - n\mu)/\sqrt{n}.$$

Of course with (1.4) we have that

$$E(S_n^*) = 0, \qquad \text{var}(S_n^*) = 1. \tag{1.5}$$

However, the crucial aspect of the denominator of (1.4) is not the achievement of unit standard deviation but the achievement of asymptotically constant standard deviation. Indeed the version omitting σ in the denominator turns out to be particularly advantageous in the multivariate case. We shall refer to all such variables as standardized, leaving it to be understood from the context whether standardization to unit variance is involved.

If we were working with random variables whose first two moments did not exist, we would have to use a different method of standardization.

Suppose now that the Y_j's, while independent, are not identically distributed; we write

$$E(Y_j) = \mu^{(j)}, \qquad \operatorname{var}(Y_j) = \kappa^{(j)} = \sigma_j^2. \tag{1.6}$$

Then with S_n again defined by (1.2), we have that

$$E(S_n) = \sum_{j=1}^{n} \mu^{(j)}, \qquad \operatorname{var}(S_n) = \sum_{j=1}^{n} \kappa^{(j)},$$

say. It is thus natural to replace (1.4) by

$$S_n^* = \left(S_n - \sum_{j=1}^{n} \mu^{(j)}\right) \Big/ \sqrt{\sum_{j=1}^{n} \kappa^{(j)}}, \tag{1.7}$$

so that (1.5) is again satisfied. It may, however, be more convenient to replace the standardizing constants by asymptotic equivalents. For example, suppose that for suitable constants b_μ, $b_\kappa = 2b_\sigma > 0$, $\kappa > 0$ and μ, we have

$$\sum_{j=1}^{n} \mu^{(j)} \sim \mu n^{b_\mu}, \qquad \sum_{j=1}^{n} \kappa^{(j)} \sim \kappa n^{b_\kappa} = \sigma^2 n^{2b_\sigma}, \tag{1.8}$$

where the sign $\sim$ is used in the sense that, for example, as $n \to \infty$

$$\lim \sum_{j=1}^{n} \mu^{(j)}/(\mu n^{b_\mu}) = 1.$$

Then it may well be an advantage to replace (1.7) by

$$S_n^{*'} = (S_n - \mu n^{b_\mu})/(\sigma n^{b_\sigma}) \tag{1.9}$$

for which (1.5) is satisfied only asymptotically. Considerations of simplicity apart, the best choice of standardizing constants in any particular application depends on the precise objective of the approximation; the exact achievement of (1.5) may have no special merit.

Note that if the originating random variable is subject to a linear transformation then the standardized form (1.4) is unchanged. For example, if instead of S_n we work with the mean $\bar{Y}_n = S_n/n$, we have that

$$\bar{Y}_n^* = (\bar{Y}_n - \mu)(\sigma/\sqrt{n})^{-1} = S_n^*.$$

Example 1.1 Linear regression

Let $X_1, X_2, \ldots$ be independent random variables and let $z_1, z_2, \ldots$ be constants such that

$$E(X_j) = \beta z_j, \qquad \text{var}(X_j) = \sigma^2 \qquad (j = 1, 2, \ldots).$$

The least squares estimate of β from $\{X_1, \ldots, X_n\}$ is

$$\hat{\beta}_n = \sum(z_j X_j)/\sum z_j^2. \tag{1.10}$$

Let $Y_j = z_j X_j$ so that $\mu^{(j)} = \beta z_j^2$, $\kappa^{(j)} = z_j^2\sigma^2$ and the standardized form of $S_n = Y_1 + \cdots + Y_n$ given by (1.7) is

$$\frac{S_n - \beta\sum z_j^2}{\sigma\sqrt{\sum z_j^2}}$$

which is, of course, the form obtained by standardizing $\hat{\beta}_n$ directly. With some relatively minor amendments the discussion applies also to the least squares estimate of a component parameter in a more general linear model.

If, in particular, $z_j = jh$, so that the values of the 'explanatory variable' are equally spaced, then

$$\sum_{j=1}^{n} z_j^2 = h^2 \sum_{j=1}^{n} j^2 \sim \tfrac{1}{3}h^2 n^3,$$

exemplifying (1.8) with $b_\mu = 3$, $b_\kappa = 3$, $b_\sigma = 3/2$.

Finally, if the originating sequence $\{Y_1, Y_2, \ldots\}$ consists of non-independent random variables, study of S_n will naturally start by examination of the exact or asymptotic form of $E(S_n)$ and $\text{var}(S_n)$. See Example 1.5.

Note that standardization is by linear transformation. For continuous random variables it would be fruitless to allow arbitrary nonlinear standardizing transformations because any continuous distribution can be transformed into any other. In special applications, however, it may be very useful to consider simple families of nonlinear transformations of the random variables of interest, such as powers and logs. We shall have examples later.

1.3 Moments, cumulants and their generating functions

We now return to the simplest case of independent and identically distributed Y_j, copies of a random variable Y. We write

$$\mu_r' = E(Y^r), \qquad \mu_r = E(Y - \mu)^r \qquad (r = 1, 2, \ldots) \tag{1.11}$$

for the moments of Y about the origin and about the mean, $\mu_1' = \mu$; we write also $\mu_2 = \kappa = \sigma^2$. If we need to emphasize the random variable Y we write $\mu(Y)$, etc.

Next we define the *moment generating function* of Y by $M(Y;t) = M(t) = E(e^{tY})$. Note our general convention of notation whereby the random variable under consideration is to the left of the semicolon and is suppressed when clear from the context. We often assume that $M(t)$ is convergent in an interval of real values of t containing $t = 0$ in its interior, so that we have the expansions

$$M(Y;t) = 1 + \sum \mu_r' t^r / r!, \tag{1.12}$$

$$M(Y - \mu; t) = e^{-t\mu} M(Y;t) = 1 + \sum \mu_r t^r / r!. \tag{1.13}$$

Recall that if Y is normally distributed with mean μ and variance σ^2, which we write $N(\mu, \sigma^2)$, then

$$M(Y;t) = \exp(t\mu + \tfrac{1}{2}t^2\sigma^2), \qquad M(Y - \mu; t) = \exp(\tfrac{1}{2}t^2\sigma^2). \tag{1.14}$$

Many other special cases are subsumed in the following general result for the exponential family.

Example 1.2 Exponential family

Suppose that X has the density

$$\exp\{x\theta - k(\theta) + a(x)\}. \tag{1.15}$$

Now the normalizing condition

$$\int \exp\{x\theta - k(\theta) + a(x)\} dx = 1$$

implies that for all θ

$$\begin{aligned} M(t) &= \int \exp\{xt + x\theta - k(\theta) + a(x)\}dx \\ &= \exp\{k(\theta + t) - k(\theta)\}. \end{aligned} \tag{1.16}$$

The normal distribution $N(\mu, \sigma_0^2)$, for fixed σ_0^2, is the special case $\theta = \mu/\sigma_0^2$,

$$k(\theta) = \tfrac{1}{2}\mu^2/\sigma_0^2 = \tfrac{1}{2}\theta^2\sigma_0^2,$$

recovering the standard result.

Of course for discrete distributions we use summation (integration with respect to counting measure) in the above expressions.

Example 1.3 The log normal distribution

Let $\log Y$ have the distribution $N(\mu, \sigma^2)$, so that Y has the log normal distribution. Then

$$\mu_r'(Y) = E(Y^r) = M(\log Y; r) = \exp(r\mu + \tfrac{1}{2}r^2\sigma^2)$$

exists for all r, but the series (1.12) for the moment generating function $M(Y;t)$ is divergent for all real $t \neq 0$. As a function of the complex variable t, $M(Y;t)$ has an isolated essential singularity at the origin. This shows that our analytical assumptions about $M(Y;t)$ are rather stronger than requiring the existence of all moments and are connected with the nonuniqueness of the distribution corresponding to log normal moments.

It is convenient for many purposes to define the *cumulant generating function* $K(Y;t)$ of the random variable Y by

$$K(Y;t) = \log M(Y;t), \tag{1.17}$$

noting from (1.13) that

$$K(Y - \mu; t) = -t\mu + K(Y;t). \tag{1.18}$$

The rth cumulant κ_r of Y is defined as the coefficient of $t^r/r!$ in the Taylor expansion of (1.17), i.e. by (1.18)

$$\begin{aligned} \sum \kappa_r t^r/r! &= \log(1 + \sum \mu_r' t^r/r!) \\ &= t\mu + \log(1 + \sum \mu_r t^r/r!). \end{aligned} \tag{1.19}$$

Thus

$$\kappa_1 = \mu, \quad \kappa_2 = \mu_2 = \sigma^2, \quad \kappa_3 = \mu_3, \quad \kappa_4 = \mu_4 - 3\mu_2^2,$$
$$\kappa_5 = \mu_5 - 10\mu_3\mu_2, \quad \kappa_6 = \mu_6 - 15\mu_4\mu_2 - 10\mu_3^2 + 30\mu_2^3.$$

The inverse relations are easily written down either directly or from the expressions above. Thus

$$\mu_4 = \kappa_4 + 3\kappa_2^2, \quad \mu_5 = \kappa_5 + 10\kappa_3\kappa_2,$$
$$\mu_6 = \kappa_6 + 15\kappa_4\kappa_2 + 10\kappa_3^2 + 15\kappa_2^3.$$

By (1.18) for $r > 1$ the κ_r are invariant under the addition of a constant to Y and hence are functions of moments abou. .he mean. Note also that for any constant a,

$$\kappa_r(aY) = a^r\kappa_r(Y).$$

Further, κ_r is a function of the moments only up to order r and hence can be defined regardless of the convergence of moments of order higher than r.

Note from the form (1.14) for the moment generating function of the normal distribution that the corresponding cumulant generating function is

$$\mu t + \tfrac{1}{2}\sigma^2 t^2,$$

so that

$$\kappa_1 = \mu, \quad \kappa_2 = \sigma^2, \quad \kappa_r = 0 \qquad (r = 3, \ldots). \tag{1.20}$$

More generally it follows from (1.16) that the cumulant generating function of the exponential family density (1.15) is $k(\theta + t) - k(t)$. We call $k(\theta)$ the *cumulant transform* of the family.

It is often helpful to write the cumulants in dimensionless form by introducing the standardized cumulants

$$\rho_r = \kappa_r/\sigma^r \qquad (r = 3, \ldots)$$

and in particular

$$\rho_3 = \kappa_3/\sigma^3, \qquad \rho_4 = \kappa_4/\sigma^4. \tag{1.21}$$

Note that ρ_3, ρ_4 are often denoted by γ_1, γ_2 or in an older notation by $\sqrt{\beta_1}$ and $\beta_2 - 3$. They are called respectively measures of skewness and kurtosis. For symmetrical distributions $\rho_3 = 0$ and indeed all the odd-order cumulants vanish, provided they exist.

Moments and cumulants, especially the first four, are used in applications in a number of rather different ways. The mean, variance and standardized third and fourth cumulants stand in their own right as measures of location, scatter, symmetry and long-tailedness, provided that the distribution concerned is reasonably well behaved in the extreme tails; all, especially the last two, are, of course, very dependent on the tails. We shall see in Chapter 4 that the standardized cumulants arise naturally in studying asymptotic expansions for the distribution of sums of independent random variables. In a more empirical way, if all we know about a distribution is its first four cumulants, it can be helpful to find a smooth distribution with those cumulants and to use that as an approximation; this is appropriate even when the cumulants are not of intrinsic interest. The family of Pearson curves is often used in this way via Pearson and Hartley (1966, Table 42). Finally we may use cumulants to assess closeness of one parametric distributional form to another and of empirical data to particular distributional forms, although in the latter case estimation errors in the higher cumulants are likely to be substantial.

For example, by (1.20), closeness to the normal distribution can be assessed by the values of the higher cumulants. The cumulants can be used similarly to construct comparisons with any other standard distribution for which a relation between cumulants is known. We illustrate comparison with a gamma distribution.

Example 1.4 Gamma distribution

The density of the gamma distribution can be parameterized in various forms. For most statistical purposes the mean and the index are particularly convenient parameters, but here we shall use the algebraically rather simpler form

$$\lambda(\lambda x)^{\beta-1}e^{-\lambda x}/\Gamma(\beta)$$

from which the cumulant generating function is easily calculated as

$$-\beta\log(1-t/\lambda), \tag{1.22}$$

so that, on expansion, we have

$$\kappa_1=\mu=\beta/\lambda,\quad \kappa_2=\sigma^2=\beta/\lambda^2,\quad \kappa_3=2\beta/\lambda^3. \tag{1.23}$$

This suggests defining

$$(\kappa_3/\sigma^3)(2\sigma/\mu)^{-1}$$

as an index taking the value 1 for any gamma distribution and, for example exceeding one if the distribution has a higher third cumulant than the gamma distribution with the same coefficient of variation, σ/μ.

Now knowledge of the moments up to a given order is equivalent to that of the corresponding cumulants. Superficially the moments may seem to provide the more attractive specification because of their direct physical or geometric interpretation. Yet there are a number of respects in which, perhaps surprisingly, cumulants have a substantial advantage. This will emerge as the argument develops, but some key points are the vanishing of the cumulants for the normal distribution, their behaviour for sums of independent random variables and, especially in the multivariate case, their behaviour under linear transformation of the random variables concerned.

1.4 Properties of sums of independent random variables

We now apply the ideas of section 1.3 to the study of sums, S_n, of independent random variables. The exact distribution of S_n is given by a convolution sum or integral: for independent and identically distributed copies of a random variable Y in (1.2), continuously distributed with density $f(Y; y)$, the density $f(S_n; s)$ of S_n is given by the $(n-1)$-dimensional integral

$$f(S_n; s) = \int f(Y; y_1) \cdots f(Y; y_{n-1}) f(Y; s - y_1 - \cdots - y_{n-1}) dy_1 \cdots dy_{n-1} \tag{1.24}$$

with a corresponding sum in the discrete case. Except for very small values of n, however, (1.24) is of little use either for analytical or indeed for numerical calculation.

While numerical aspects are rather outside the scope of the present book, it is worth noting that the most effective numerical methods for the evaluation of (1.24) depend on finding a density $f_0(y)$ for which the convolutions can be calculated explicitly in simple form, and then writing either

$$f(Y; y) = f_0(y) + \delta(y),$$

where $\delta(y)$ is a small perturbation, or

$$f(Y; y) = f_0(y)\{1 + \sum c_r p_r(y)\},$$

where $\{p_r(y)\}$ is the set of orthogonal polynomials associated with $f_0(y)$; see section 1.6. In both cases a series expansion is obtained for the required convolution, in which the leading terms are relatively easy to compute and in which it may be possible to arrange that the leading terms are of primary importance. The normal, gamma, Poisson and binomial are the main candidates for $f_0(y)$.

For theoretical development, however, it is more fruitful to argue in terms of moment generating functions, in view of relations which for independent and identically distributed components become

$$M(S_n; t) = E(e^{tS_n}) = \prod E(e^{tY_j}) \tag{1.25}$$

$$= \{M(Y; t)\}^n, \tag{1.26}$$

as could also be proved analytically by applying a Laplace transformation to (1.24). Thus for the cumulant generating function we have that

$$K(S_n; t) = nK(Y; t), \tag{1.27}$$

so that on expansion

$$\kappa_r(S_n) = n\kappa_r(Y) \qquad (r = 1, 2, \ldots), \tag{1.28}$$

generalizing (1.3). The simplicity of (1.28) is one important reason for introducing cumulants; note that in general $\mu_r(S_n) \neq n\mu_r(Y)$ $(r > 3)$. Because of invariance under linear transformation, $\rho_r(S_n^*) = \rho_r(S_n)$ $(r = 3, \ldots)$. Also from (1.28) there follows immediately the important result that

$$\rho_3(S_n) = \rho_3(Y)/\sqrt{n}, \quad \rho_4(S_n) = \rho_4(Y)/n, \quad \rho_r(S_n) = \rho_r(Y)/n^{(1/2)r-1}. \tag{1.29}$$

Now suppose, as in (1.6) and (1.7), that the random variables $\{Y_1, Y_2, \ldots\}$, while independent, are not identically distributed. Equation (1.25) still holds, (1.26) being replaced by

$$M(S_n; t) = \prod M(Y_j; t) \tag{1.30}$$

and (1.27) by

$$K(S_n; t) = \sum K(Y_j; t). \tag{1.31}$$

Thus (1.21) for the standardized skewness and kurtosis becomes

$$\rho_3(S_n) = \sum \kappa_3(Y_j)/\{\sum \kappa_2(Y_j)\}^{3/2},$$

$$\rho_4(S_n) = \sum \kappa_4(Y_j)/\{\sum \kappa_2(Y_j)\}^2,$$

and the dependence on n could take many forms. If, however, the

Y_j all have the same ρ_3 and ρ_4, although different variances, then

$$\rho_3(S_n) = \rho_3 \sum \sigma_j^3 / (\sum \sigma_j^2)^{3/2}, \tag{1.32}$$

$$\rho_4(S_n) = \rho_4 \sum \sigma_j^4 / (\sum \sigma_j^2)^2. \tag{1.33}$$

One elementary way of exploring the effect on (1.32) of unequal σ_j is to write

$$\sigma_j = \bar{\sigma}(1 + \varepsilon\delta_j),$$

where $\sum \delta_j = 0$, so that $\bar{\sigma}$ is the average of the σ_j and ε is a notional small quantity. Let $\sum \delta_j^2 / n = \mu_2(\delta)$ be the 'mean square' of the δ_j; note that $\varepsilon^2 \mu_2(\delta)$ is the squared coefficient of variation of the σ_j, $\text{cv}^2(\sigma_j)$.

Then if we take terms only to order ε^2, we have that

$$\begin{aligned} \rho_3(S_n) &= \rho_3 n^{-1/2}\{1 + 3/2\varepsilon^2 \mu_2(\delta)\} \\ &= \rho_3 n^{-1/2}\{1 + 3/2\text{cv}^2(\sigma_j)\} \end{aligned} \tag{1.34}$$

and similarly

$$\begin{aligned} \rho_4(S_n) &= \rho_4 n^{-1}\{1 + 4\varepsilon^2 \mu_2(\delta)\} \\ &= \rho_4 n^{-1}\{1 + 4\,\text{cv}^2(\sigma_j)\}. \end{aligned} \tag{1.35}$$

Thus, as is intuitively clear, and can easily be proved in generality, variation of the $\{\sigma_j\}$ increases $|\rho_3(S_n)|$ and $|\rho_4(S_n)|$ as compared with the corresponding values for equal $\{\sigma_j\}$. The simple expansions give an idea of the magnitude of the effects. As a rough but tough trial of the accuracy of such approximations, consider the special case in which $\frac{1}{2}n$ of the σ_j are zero and the other $\frac{1}{2}n$ are 2σ, so that the coefficient of variation of the $\{\sigma_j\}$ is one. Then the approximate inflation factors are 5/2 and 5 compared with the exact factors of 2 and 4, showing rough agreement even in a rather extreme case.

It is immediate that if the Y_j have all the same variance but different third and fourth cumulants, then $\rho_3(S_n)$ and $\rho_4(S_n)$ are determined by the average ρ_3 and ρ_4.

1.5 Calculation of moments and cumulants

In the previous section moments and cumulants of sums of independent random variables were calculated in terms of those of component random variables, full specification of the distributions of the components being unnecessary. Similar exact calculations can be carried out in a general way only for polynomial functions of the basic random variables. Use of special techniques is desirable if these calculations are to be done in any generality and in such an approach cumulants are the most fruitful quantities rather than moments; see McCullagh (1987) and Chapters 5 and 6 below. Here we merely illustrate some of the simpler cases and give some results to be used later.

Let $\{Y_1, Y_2, \ldots\}$ be independent and identically distributed random variables as in (1.2). If interest lies in one of

$$S_{n,r} = Y_1^r + \cdots + Y_n^r, \qquad M'_{n,r} = n^{-1}S_{n,r} \qquad (r = 1, 2, \ldots), \tag{1.36}$$

the previous discussion applies replacing Y by Y^r. Note that

$$\mathrm{var}(Y^r) = E(Y^{2r}) - \{E(Y^r)\}^2 = \mu'_{2r} - \mu_r'^2, \tag{1.37}$$

reducing to $\mu_{2r} - \mu_r^2$ if $E(Y) = 0$. Thus

$$\mathrm{var}(M'_{n,r}) = (\mu'_{2r} - \mu_r'^2)/n, \tag{1.38}$$

etc. More interest usually attaches to moments about the mean

$$M_{n,r} = n^{-1}\{(Y_1 - \bar{Y}_n)^r + \cdots + (Y_n - \bar{Y}_n)^r\}, \tag{1.39}$$

where $\bar{Y}_n = M'_{n1} = S_n/n = (Y_1 + \cdots + Y_n)/n$, and in particular to $M_{n,2}$. The evaluation of the moments of $M_{n,2}$ proceeds most directly by supposing, without loss of generality, that $E(Y) = 0$, and then writing

$$M_{n,2} = n^{-1}\{\textstyle\sum Y_j^2(1 - 1/n) - 2n^{-1}\sum Y_i Y_j\},$$
$$M_{n,2}^2 = n^{-2}\{\textstyle\sum Y_j^4(1 - 1/n)^2 + 2\sum Y_i^2 Y_j^2[(1 - 1/n)^2 + 2/n^2]\} + \eta,$$

where η is a collection of terms involving at least one Y_j to unit power, and hence having zero expectation. Thus

$$E(M_{n,2}) = \sigma^2(n-1)/n,$$
$$E(M_{n,2}^2) = \mu_4(n-1)^2/n^3 + \sigma^4(n-1)(n^2 - 2n + 3)/n^3,$$
$$\mathrm{var}(M_{n,2}) = \mu_4(n-1)^2/n^3 - \sigma^4(n-1)(n-3)/n^3. \tag{1.40}$$

Note that for large n

$$\mathrm{var}(M_{n,2}) \doteq \mathrm{var}(M'_{n,2}) \doteq (\mu_4 - \sigma^4)/n; \tag{1.41}$$

the reason for this will appear later. Similarly

$$\mathrm{cov}(M_{n,2}, \bar{Y}_n) \doteq \mu_3(n-1)/n^2.$$

Thus for large values of n

$$\operatorname{corr}(M_{n,2}, \bar{Y}_n) \doteq \rho_3/(2 + \rho_4). \tag{1.42}$$

Further results are recorded in the exercises. Incidentally, note from the cubic structure of $\operatorname{cov}(M_{n,2}, \bar{Y}_n)$ that it must have the form $a_n\mu_3 + b_n\mu\mu_2$ and that, because of invariance under translation, $b_n = 0$.

Obviously if one wished to pursue these calculations for higher moments some special more economical methods of calculations would be desirable.

Similar calculations can be made for polynomials in dependent random variables provided, of course, that the mixed moments $E(Y_1 Y_2 \cdots)$ up to appropriate order are specified. Two special cases of particular interest concern dependencies associated with stationary time series models and those connected with random sampling from a finite population. We now deal briefly with these.

Example 1.5 Stationary time series

Let $\{Y_1, \ldots, Y_n\}$ be a second-order stationary time series of zero mean, i.e. let $E(Y_i) = 0$, $E(Y_i Y_{i+h}) = \gamma(h)$, the autocovariance function, with $\operatorname{var}(Y_i) = \gamma(0)$. If $S_n = Y_1 + \cdots + Y_n$, a direct calculation shows that

$$\operatorname{var}(S_n) = \gamma(0) + 2 \sum_{h=1}^{n-1} (n - h)\gamma(h). \tag{1.43}$$

To determine an appropriate standardization we need the asymptotic behaviour of $\operatorname{var}(S_n)$ as $n \to \infty$. We shall not explore this in detail, but in fact there are essentially two cases. For processes with a finite spectral density at the origin and in which $\sum\gamma(h)$ is therefore convergent, which may be called processes with short-range dependence, $\operatorname{var}(S_n)$ is asymptotically proportional to n and standardization of S_n by division by $\sqrt{n}$ is appropriate. The other main possibility is that $\gamma(h)$ is for large h of the form ah^{-b} for $0 < b < 1$; such a process has long-range dependence, an infinite spectral density at the origin and a property of asymptotic second-order self-similarity; see, for example, Cox (1984). Here $\operatorname{var}(S_n)$ is asymptotically proportional to n^{2-b} and standardization by division by $n^{1-(1/2)b}$ is called for.

Example 1.6 Finite populations

Suppose that there is a finite population $\mathscr{Z} = \{z_1, \ldots, z_N\}$ of real numbers, usually regarded as constants not as random variables, labelled by the integers $1, \ldots, N$. From this a sample of size n is drawn randomly without replacement giving rise to random variables $Y_1, \ldots, Y_n$. The term randomly without replacement can be defined mathematically in various ways of which the following two are the most important:

1. The sample space consists of all distinct sets of n individuals chosen from $\mathscr{Z}$, without regard to order, and each of these sets (possible samples) is assigned probability $1/\binom{N}{n}$;
2. The first random variable Y_1 is defined by choosing a label equally likely to be any one of $\{1, \ldots, N\}$ and then defining Y_1 to be the corresponding z. Given Y_1, the random variable Y_2 is defined similarly by random choice from the remaining labels, and so on.

It is easy to show that (1) and (2) are compatible; for (1), in order to define the individual Y_j it is necessary to randomize the order within each sample, but for many purposes the individual Y_j are not required, because the properties of the sample without regard to order are relevant. In this respect (1) is a simpler definition.

Note that (2) implies probability statements such as

$$P(Y_j = z_r) = 1/N \qquad (j = 1, \ldots, n;\ r = 1, \ldots, N),$$

$$P(Y_i = z_r, Y_j = z_s) = 1/\{N(N-1)\} \qquad (i \neq j;\ r \neq s).$$

These relations can be most neatly expressed via indicator random variables

$$I_{jr} = \begin{cases} 1 & \text{if } Y_j = z_r, \\ 0 & \text{if } Y_j \neq z_r. \end{cases} \tag{1.44}$$

The above probability statements are then equivalent to a specification of the second moment properties of the indicator random variables.

Sample moments can be defined exactly as before. In particular the sample mean is $\bar{Y}_n = \sum Y_j/n$; of course the Y_j are no longer independent random variables. One way of evaluating the properties of the sample mean is to write

$$\bar{Y}_n = (\sum\sum z_r I_{jr})/n \tag{1.45}$$

and then to find the moments of low order via those of the indicator random variables.

We shall however, argue less directly. It is clear that $E(\bar{Y}_n)$ is a linear combination of the $\{z_r\}$ invariant under permutations of the labels, i.e. it is a symmetric function of degree one. Thus

$$E(\bar{Y}_n) = a_1 \sum z_r,$$

where a_1 is a constant, i.e. a function of n, N not depending on the $\{z_r\}$. Consideration of the special population $z_1 = \cdots = z_n = 1$ shows that in fact $a_1 = N^{-1}$ so that

$$E(\bar{Y}_n) = \bar{z}_N = \sum z_r/N. \tag{1.46}$$

Of course this follows also very quickly from (1.44) and (1.45).

To calculate $\operatorname{var}(\bar{Y}_n)$, we argue similarly. The answer must be of second degree in the $\{z_r\}$, be invariant under permutation of the labels and must vanish if all the z_r are equal. It follows that

$$\operatorname{var}(\bar{Y}_n) = a_2 \sum (z_r - \bar{z}_N)^2, \tag{1.47}$$

where a_2 is a constant depending only on n, N. Probably the most direct route to the calculation of a_2 is consideration of the special population $\{1, 0, \ldots, 0\}$ for which it is easily shown that

$$P(\bar{Y}_n = n^{-1}) = n/N, \qquad P(\bar{Y}_n = 0) = 1 - n/N,$$

so that a_2 can be determined by applying (1.47) to this special population, yielding

$$\begin{aligned} \operatorname{var}(\bar{Y}_n) &= \frac{N-n}{Nn} \sum \frac{(z_r - \bar{z}_N)^2}{N-1} \\ &= (1 - n/N)\kappa_{2,z}/n, \end{aligned} \tag{1.48}$$

say, where $(1 - n/N)$ is called a finite population correction and $\kappa_{2,z}$ is a second-order measure of variability of the finite population $\mathscr{Z}$. Note that if the Y_j are obtained by random sampling with replacement, i.e. are independent and identically distributed random variables each equally likely to be $\{z_1, \ldots, z_N\}$, then

$$n \operatorname{var}(\bar{Y}_n) = (N-1)\kappa_{2,z}/N. \tag{1.49}$$

A similar argument applies for the third- and higher-order cumulants and after appreciable calculation we have that under random sampling without replacement the standardized third cumulant of

$\bar{Y}_n$ is

$$(1-n/N)^{-1/2}(1-2n/N)n^{-1/2}(1-1/N)^{1/2}(1-2/N)^{-1}(m_{3,z}/m_{2,z}^{3/2}),$$

where $m_{3,z}=\sum(z_r-\bar{z}_N)^3/N$.

Theorems analogous to the classical central limit theorem are available for $\bar{Y}_n$ as $n, N\rightarrow\infty$ with n/N fixed.

The arguments sketched here for random sampling without replacement can be extended to apply both to more complex schemes of sampling, such as two-stage sampling, and to the study of randomization distributions arising in the context of experimental design. The key idea in all cases is to exploit the symmetries inherent in the problem to establish the general form of the moments under study. This is followed either by evaluation for simple special cases, as above, or, in a more subtle version of the same idea, by establishing a link with known behaviour for independent and identically distributed random variables.

1.6 Appendix: Orthogonal polynomials

In various detailed calculations it is convenient to use orthogonal polynomials. Such polynomials have an extensive literature and are of interest in a number of mathematical contexts. Here we outline a few key ideas in the one-dimensional case using a probabilistic terminology; for the multidimensional case, see section 5.7.

Suppose we are given a univariate distribution, discrete or continuous, all of whose moments exist. Then $\{p_k(x)\}$ for $k=0,1,\ldots$ are orthogonal polynomials with respect to that distribution if

(i) $p_k(x)$ is a polynomial of degree k;
(ii) $E\{p_k(X)p_l(X)\}=0 \quad (k\neq l=0,1,\ldots)$,

the expectation being with respect to the given distribution. As so defined the polynomials are undetermined to constant multipliers. We can make them unique in various ways, for example by normalizing in mean square,

(iii) $E\{p_k(X)^2\}=1$ with leading coefficient positive; in particular $p_0(x)=1$,

or by simplifying the leading term by requiring that

(iii)′ the coefficient of x^k in $p_k(x)$ is 1; in particular $p_0(x)=1$.

It will be clear from the context whether (iii) or (iii)′ or indeed some other convention is in use. Note that $E\{p_k(X)\} = 0$ $(k = 1, 2, \ldots)$.

Orthogonal polynomials are in principle easily constructed by Gram–Schmidt orthogonalization of the sequence $\{1, x, x^2, \ldots\}$. Thus we write

$$p_k(x) = x^k + a_{k1}x^{k-1} + \cdots + a_{kk}$$

and determine the a_{kj} by the orthogonality condition (ii) for $l = 0, \ldots, k-1$. The coefficients can be expressed in terms of the moments (or cumulants) of the defining distribution. Thus

$$p_1(x) = x - \mu, \qquad p_2(x) = (x-\mu)^2 - \mu_3(x-\mu)/\mu_2 - \mu_2,$$

etc., with alternative expressions in terms of moments about the origin.

Now any polynomial, $q_l(x)$, of degree l can be written as a linear combination of $p_0(x), \ldots, p_l(x)$ and it follows immediately from (ii) that $E\{p_k(X)q_l(X)\} = 0$ $(l < k)$.

To expand an arbitrary function $h(x)$ we write formally

$$h(x) = \sum h_l p_l(x). \tag{1.50}$$

On multiplying by $p_k(x)$ and taking expectations, we have with the normalization (iii),

$$h_k = E\{h(X)p_k(X)\}.$$

Conditions of completeness are needed for the expansion to be identically true; in general, however, the expansion truncated at any finite point gives a least squares approximation to the function $h(x)$.

Sometimes it may be more sensible to write

$$h(x) = \sum h_l' p_l(x) f(x), \tag{1.51}$$

where $f(x)$ is the density of the defining distribution, i.e. in effect to apply the previous argument to $h(x)/f(x)$. Then

$$h_k' = \int h(x)p_k(x)dx, \tag{1.52}$$

where the integral or sum is over the support of the distribution $f(x)$. Expansions (1.50) and (1.51) are, however, to be used, if at all, with considerable caution. In particular the conditions for convergence of the infinite series are strong. Thus if $f(x) = \phi(x)$, the standard normal density, and $h(x)$ is a probability density then convergence of (1.51) requires $e^{1/4x^2}h(x)$ to have convergent integral.

Important general properties of orthogonal polynomials include their having distinct real zeros and their obeying three-term recurrence equations. The so-called classical orthogonal polynomials are those associated with normal, gamma and beta distributions and have many further properties.

We deal briefly with the Hermite polynomials associated with the normal distribution and the Laguerre polynomials associated with the gamma distribution.

To study the Hermite polynomials, suppose that the defining distribution is normal with zero mean and standard deviation σ. Then $p_k(x) = H_k(x/\sigma)$, where $H_k(x)$ is the polynomial for unit standard deviation. This rather trivial transformation law, an immediate consequence of the defining integral, becomes much more significant in the multidimensional case. The most compact definition of $H_k(x)$ is

$$\phi(x)H_k(x) = (-1)^k(d/dx)^k\phi(x).$$

It is immediate that this defines a polynomial of degree k with leading coefficient one; the orthogonality can be proved by integration by parts.

The first few Hermite polynomials are

$$H_0(x) = 1, \quad H_1(x) = x, \quad H_2(x) = x^2 - 1,$$
$$H_3(x) = x^3 - 3x, \quad H_4(x) = x^4 - 6x^2 + 3,$$
$$H_5(x) = x^5 - 10x^3 + 15x, \quad H_6(x) = x^6 - 15x^4 + 45x^2 - 15.$$

The Hermite polynomials have a simple generating function

$$\sum_{k=0}^{\infty} z^k H_k(x)/k! = \exp(xz - \tfrac{1}{2}z^2) \tag{1.53}$$

from which further properties can be deduced. Another property of direct relevance to moment generating functions is

$$\begin{aligned}\int e^{tx}H_k(x)\phi(x)dx &= E\{e^{tX}H_k(X)\}\\ &= t^k \exp(\tfrac{1}{2}t^2).\end{aligned}$$

The generalized Laguerre polynomials are defined somewhat

analogously with respect to a gamma distribution $x^{\alpha-1}e^{-x}/\Gamma(\alpha)$

$$L_k^{\alpha}(x) = (-1)^k e^x x^{-\alpha+1}(d/dx)^k(e^{-x}x^{k+\alpha-1}).$$

Properties are studied much as for the Hermite polynomials. The most important special case has $\alpha = 1$ giving the ordinary Laguerre polynomials orthogonal with respect to an exponential distribution. The first few are

$$L_0(x) = 1, \quad L_1(x) = x - 1, \quad L_2(x) = x^2 - 4x + 2,$$
$$L_3(x) = x^3 - 9x^2 + 18x - 6.$$

In statistical applications these arise mostly when a limiting distribution is exponential as in some extreme value problems and in point process theory.

Further results and exercises

1.1 When the moment generating function $M(Y;t) = E(e^{tY})$ of a random variable Y does not converge for real t in an interval containing the origin, it is standard to work with the characteristic function $E(e^{itY})$ which always exists for real t, and which uniquely determines the distribution; it obviously has the same formal moment generating property as $M(Y;t)$. That is, one works with Fourier transforms rather than Laplace transforms. If moments of all orders exist and if the series $\sum \mu_r' z^r/r!$ as a function of the complex variable z has positive radius of convergence then the moments determine the characteristic function and hence the distribution.

(a) Show that the characteristic function of the Cauchy density $\{\pi(1+y^2)\}^{-1}$ is $e^{-|t|}$ and comment on the connexion between the form of this and the nonexistence of the moments.
(b) Show that the Poisson distribution of mean μ has all its cumulants equal to μ and that the Poisson distribution is the only distribution with this property.
(c) Note the connexion with the discussion of the log normal distribution in Example 1.3.

[Section 1.3]

1.2 Obtain for an arbitrary random variable Y

(a) $\text{var}\{(Y-\mu)^2\}$,
(b) $\text{cov}\{Y-\mu, (Y-\mu)^2\}$.

By noting that (a) must be nonnegative and applying the Cauchy–Schwarz inequality to the second and expressing the results in terms of standardized cumulants, prove that $\rho_4 \geqslant -2$ and that $\rho_4 - \rho_3^2 + 2 \geqslant 0$. When is equality achieved?

[Section 1.3]

1.3 For an arbitrary collection of m random variables $(Y_1, \ldots, Y_m)$ the covariance matrix is defined to be the $m \times m$ symmetric matrix with (i, j)th element $\text{cov}(Y_i, Y_j)$. Such matrices are positive semidefinite always and positive definite if there are no linear relations holding with probability one. Recover the results of exercise 1.2 via the covariance matrix of $(Y - \mu, (Y - \mu)^2)$. Generalize by considering the covariance matrix of $(Y - \mu, \ldots, (Y - \mu)^p)$.

[Section 1.3]

1.4 For two positive integers m and r define the rth descending factorial power of m as $m^{(r)} = m(m-1)\cdots(m-r+1)$. Ordinary and factorial powers are connected via the relations

$$m^r = \sum(\Delta^j 0^r/j!)m^{(j)},$$

where the $(\Delta^j 0^r/j!)$ are the Stirling numbers of the second kind. Factorial moments are defined by $\mu_{(r)} = E(Y^{(r)})$.

(a) Obtain the factorial moments of the Poisson distribution.
(b) Obtain the connexion between factorial moments and the probability generating function.
(c) Define in a suitable way factorial cumulants.

[Section 1.3]

1.5 Note that it is formally possible to define moments of negative, fractional and indeed arbitrary order, in general by the Mellin transform $E(Y^t)$ considered for real t. Obtain the conditions for the existence of negative moments of the gamma distribution, in particular the condition for the existence of the expected reciprocal.

[Section 1.3]

1.6 Let $f_1(x)$, $f_2(x)$ be two probability density functions of zero mean and unit variance and third and fourth cumulants $\kappa_3^{(1)}, \kappa_4^{(1)}$ and $\kappa_3^{(2)}$, $\kappa_4^{(2)}$ respectively. Then a sufficient condition that $\kappa_4^{(1)} \geqslant \kappa_4^{(2)}$ is that: (i) there exist four abscissae $a_1 < a_2 < a_3 < a_4$ such that $f_1 \geqslant f_2$

in $(-\infty < x < a_1) \cup (a_2 < x < a_3) \cup (a_4 < x < \infty)$ and that $f_1 \leqslant f_2$ in $(a_1 < x < a_2) \cup (a_3 < x < a_4)$; and (ii) $a_1 + a_2 + a_3 + a_4$ and $\kappa_3^{(2)} - \kappa_3^{(1)}$ are not both strictly positive or strictly negative. To prove this, note that

$$\int (x - a_1)(x - a_2)(x - a_3)(x - a_4)(f_1 - f_2)dx \geqslant 0.$$

Show further that the condition is not necessary even when both densities are unimodal and with a single inflexion on each side of the mode. For this take $f_1 = \phi(x)$ and

$$f_2(x) = \phi(x)\{1 + A(x^8 - 22x^6 + 116x^4 - 126x^2 + 3)\},$$

where $\phi(x)$ is the standard normal density and A is a small positive constant.

[Section 1.3; Dyson, 1943]

1.7 Consider a finite population of size N consisting of R 1's and $N - R$ 0's. A random sample of size n is drawn without replacement; write down the hypergeometric distribution governing the number of 1's in the sample. Obtain the mean and variance of the distribution. Obtain the mean and variance also directly from the general formulae (1.46) and (1.48), noting the much greater simplicity of the second approach.

[Section 1.5]

1.8 Examine the special case of exercise 1.7 in which $R = 1$, obtaining the standardized third cumulant of the distribution of the sample total (or mean). Compare it with the value for random sampling with replacement, i.e. for the binomial distribution with probability of 'success' $1/N$.

[Section 1.5]

1.9 Obtain $\text{corr}\{M_{n,2}, (\bar{Y}_n - \mu)^2\}$ and comment on the possible statistical uses of this and (1.42).

[Section 1.5]

1.10 Orthogonal polynomials can be associated with the beta density on (0, 1), namely $x^a(1 - x)^b / B(a + 1, b + 1)$. It is convenient to work instead with a density on $(-1, 1)$ proportional to $(1 - x)^a$

$(1+x)^b$. The resulting orthogonal polynomials, denoted by $P_n^{(a,b)}(x)$, are called Jacobi polynomials. It can be shown that

$$P_n^{(a,b)}(x) = (-1)^n \frac{(1-x)^{-a}(1+x)^{-b}}{2^n n!} \frac{d^n}{dx^n}\{(1-x)^{n+a}(1+x)^{n+b}\}.$$

Verify the orthogonality property of this form by repeated integration by parts.

[Section 1.6]

1.11 Prove the result (1.53) giving the generating function for Hermite polynomials and from it deduce a differential-difference equation expressing $H_{k+1}(x)$ in terms of $H_k'(x)$ and $H_k(x)$. Hence obtain the first six polynomials starting from $H_0(x) = 1$. More generally, the Meixner polynomials are defined via a generating function

$$g(z)\exp\{xh(z)\},$$

where $g(z)$, $h(z)$ have power series expansions in z. Show that this family includes in addition to the normal, the gamma, Poisson, binomial and negative binomial distributions. The generating functions for the last four are

$$(1+z)^{-\alpha}\exp\{xz/(1+z)\},\ (1+z)^x e^{\mu z},$$
$$\{1+(1-\theta)z\}^x(1-\theta z)^{n-x},\ (1+z\theta)^{-x-\alpha}\{1+z(1+\theta)\}^x.$$

[Section 1.6]

1.12 Prove from the definition of Laguerre polynomials in section 1.6 that

(a) $$L_k^\alpha(x) = \sum (k+\alpha-1)^{(k-j)k}C_j(-1)^{k-j}x^j,$$

(b) $$L_k^\alpha(x) = L_k^{\alpha+1}(x) + kL_{k-1}^{\alpha+1}(x),$$

(c) $$(d/dx)L_k^\alpha(x) = kL_{k-1}^{\alpha+1}(x),$$

(d) $$\int_0^\infty e^{tx}\frac{x^{\alpha-1}e^{-x}}{\Gamma(\alpha)}L_k^\alpha(x)dx = \frac{(k+\alpha)^{(k)}t^k}{(1-t)^{k+\alpha}},$$

this being of particular use in connexion with the inversion of moment generating functions.

[Section 1.6]

1.13 Orthogonal expansions of the form (1.51) are called Gram–Charlier series of Type A or of Type B when the base distribution is, respectively, the normal or the Poisson distribution. If $\phi(x;\mu,\sigma^2)$ is the normal density function of mean μ and variance σ^2 and if Z is a pseudo-variable of zero mean and variance and higher cumulants the same as those of X, show that formal Taylor expansion of

$$f_X(x) = E_Z\phi(x;\mu + Z, \sigma^2)$$

yields the Type A series.

[Section 1.6; Davis, 1976]

1.14 Show by the use of a generating function or otherwise that if $a_1, \ldots, a_r$ are such that $a_1^2 + \cdots + a_r^2 = 1$, then formally for Hermite polynomials

$$H_m(a_1x_1 + \cdots + a_rx_r) = \{a_1H(x_1) + \cdots + a_rH(x_r)\}^m,$$

where, when the right-hand side is expanded, $\{H(x_j)\}^n$ is to be interpreted as $H_n(x_j)$.

[Section 1.6; Kampé de Fériet, 1923]

Bibliographic notes

Much of the material in this chapter is widely available in textbooks. Cumulants were introduced by Thiele (1889, 1897, 1899) and rediscovered and much exploited by R.A. Fisher.

The exponential family was mentioned by Fisher (1934) and systematically formalized in the mid-1930s independently by G. Darmois, L.H. Koopman and E.J.G. Pitman, after some or all of whom it is sometimes named. For a thorough modern account of its properties, see Barndorff-Nielsen (1978, 1988).

The question of when a distribution is uniquely determined by its moments is classical; see, for example, Shohat and Tamarkin (1943).

The systematic use of symmetric functions in studying sampling from a finite population starts with Irwin and Kendall (1944), following an idea of R.A. Fisher. Tukey (1950) made an important extension; for a recent account, see McCullagh (1987). Kendall and Stuart (1969, Ch. 12) give many detailed formulae. For a central limit theorem for sampling from a finite population under Lindeberg type conditions, see Renyi (1970, p. 460). The closely related application in the randomization theory of experimental design can be traced to gnomic remarks of R.A. Fisher and more systematically to Yates (1951) and Grundy and Healy (1950); for a mathematically much more general account, see Bailey (1981), and Bailey and Rowley (1987).

A major analytical account of orthogonal polynomials is by Szegö (1967, 3rd ed.) and a more introductory discussion is by Chihara (1978). Many books on the methods of mathematical physics treat the matter incidentally; see particularly Olver (1974). For detailed properties of special cases, see Abramowitz and Stegun (1965). For systematic use of orthogonal expansions in a statistical context, especially in connexion with bivariate distributions, see Lancaster (1969) and the references therein.

CHAPTER 2

Some basic limiting procedures

2.1 Introduction

In the previous chapter we outlined a number of key matters, especially notions associated with sums of independent random variables. We now turn to some further important ideas, involving in particular the approximation of functions of random variables by local linearization, a procedure often referred to as the δ-method.

The central ideas are best illustrated by a simple example. This is followed by a fairly general although still quite informal account illustrated by a number of further examples. In the initial example free use is made of terms like convergence in distribution that are defined more carefully later in the chapter.

2.2 A simple example

Suppose that the random variable U_n has the distribution $\mathrm{N}(\mu, \sigma^2/n)$; for example, U_n might be the mean of n independent random variables with the distribution $\mathrm{N}(\mu, \sigma^2)$. Suppose further that we are interested in the behaviour for large n of simple functions of U_n, and $V_n = U_n^2$ and $W_n = \exp(pU_n)$, for fixed p, will serve as convenient examples. The exact distributions of V_n and W_n, respectively, being proportional to noncentral chi-squared and being log normal can, of course, be studied directly, so that we are able to compare our approximate results with the corresponding exact ones.

To examine behaviour for large n, we first standardize U_n by writing

$$U_n^* = (U_n - \mu)\sqrt{n}/\sigma, \qquad U_n = \mu + \sigma U_n^*/\sqrt{n}, \tag{2.1}$$

leading to

$$V_n = \mu^2 + 2\mu\sigma U_n^*/\sqrt{n} + \sigma^2 U_n^{*2}/n, \tag{2.2}$$

$$\begin{aligned} W_n &= e^{p\mu} \exp(p\sigma U_n^*/\sqrt{n}) \\ &= e^{p\mu}(1 + p\sigma U_n^*/\sqrt{n} + \cdots). \end{aligned} \tag{2.3}$$

We call (2.2) and (2.3) *stochastic expansions* for the random variables V_n and W_n.

Now U_n^*, which is a standard normal variable, takes, with very high probability, values in a bounded interval, say $(-4, 4)$. Thus the terms of the expansions (2.2) and (2.3) are, for sufficiently large n, in decreasing order. A number of conclusions follow:

1. First, if we note that certain terms tend to zero as $n \to \infty$, we have that $V_n \to \mu^2$ and $W_n \to e^{p\mu}$ in probability. These results are special cases of a result to be noted later that continuous functions of random variables converging in probability also converge.
2. Secondly, if we ignore terms of order $1/n$, V_n and W_n are linear functions of U_n^* and hence are approximately normally distributed with approximate means $\mu^2, e^{p\mu}$ and variances
$$4\mu^2\sigma^2/n, \quad e^{2p\mu}p^2\sigma^2/n.$$
The exact means are easily seen to be
$$\mu^2 + \sigma^2/n, \quad \exp(p\mu + \tfrac{1}{2}p^2\sigma^2/n)$$
and the variances
$$4\mu^2\sigma^2/n + 2\sigma^4/n^2,$$
$$\exp(2p\mu + p^2\sigma^2/n)\{\exp(p^2\sigma^2/n) - 1\}.$$
3. To see the approach to normality more explicitly it helps to introduce (approximately) standardized versions of V_n and W_n in the form
$$\begin{aligned} V_n^* &= (V_n - \mu^2)(2\mu\sigma/\sqrt{n})^{-1} \\ &= U_n^* + \sigma(2\mu\sqrt{n})^{-1}U_n^{*2}, \end{aligned} \tag{2.4}$$
$$\begin{aligned} W_n^* &= (W_n - e^{p\mu})(e^{p\mu}p\sigma/\sqrt{n})^{-1} \\ &= U_n^* + (\tfrac{1}{2}p\sigma/\sqrt{n})U_n^{*2} + \cdots. \end{aligned} \tag{2.5}$$
As soon as the first term is dominant, the distributions of V_n^* and W_n^* will be close to the standardized normal distribution of U_n^*.
4. If instead of having exactly a standard normal distribution, U_n^* is only asymptotically standard normal, i.e. converges in distribution to N(0, 1), then (2.4) and (2.5) show that V_n^* and W_n^*, and hence also V_n and W_n, are asymptotically normal. Of course the rate of approach to normality depends now on the properties of U_n as

well as on the later terms of (2.4) and (2.5), and particularly on how rapidly U_n itself converges to normality.

5. Supposing again that U_n is exactly $N(\mu, \sigma^2/n)$, we may obtain expansions for $E(V_n)$ and $E(W_n)$ directly from (2.2) and (2.3) as

$$E(V_n) = \mu^2 + \sigma^2/n, \qquad E(W_n) = e^{p\mu}(1 + \tfrac{1}{2}p^2\sigma^2/n + \cdots), \quad (2.6)$$

the first being, of course, exact because the underlying series expansion is finite and the second being a convergent expansion of the exact mean. In fact we shall be concerned not with the convergence of infinite series such as that for $E(W_n)$ in (2.6) but rather with the use of a limited number of terms, via consideration of (2.6) as an asymptotic expansion. For example, it is easily shown that

$$E(W_n) = e^{p\mu}\{1 + \tfrac{1}{2}p^2\sigma^2/n + O(n^{-2})\}$$

in the sense that

$$n^2[E(W_n) - e^{p\mu}\{1 + \tfrac{1}{2}p^2\sigma^2/n)\}]$$

is bounded as $n \to \infty$. For this we need a corresponding property of the expansion for W_n plus reasonable behaviour of the expectation of the omitted terms. On the whole in statistical applications, interest focuses on distributions rather than on moments as such, in particular because of an interest in the quantiles of sampling distributions.

6. To study similarly the variance or higher moments or cumulants, it is most straightforward, at least in relatively simple cases, to examine from first principles the behaviour of $\{V_n - E(V_n)\}^2$ and analogous expressions. Thus

$$V_n - E(V_n) = (2\mu\sigma/\sqrt{n})U_n^* + (\sigma^2/n)(U_n^{*2} - 1), \qquad (2.7)$$

$$\{V_n - E(V_n)\}^2 = (4\mu^2\sigma^2/n)U_n^{*2} + (4\mu\sigma^3/n^{3/2})U_n^*(U_n^{*2} - 1) + (\sigma^4/n^2)(U_n^{*2} - 1)^2. \qquad (2.8)$$

Now $E\{U_n^*(U_n^{*2} - 1)\} = 0$, $E(U_n^{*2} - 1)^2 = 2$, so that on taking expectations in (2.8), we have that

$$\text{var}(V_n) = 4\mu^2\sigma^2/n + 2\sigma^4/n^2, \qquad (2.9)$$

this in fact being exact. In the same way expressions for $\mu_r(V_n) = E\{V_n - E(V_n)\}^r$ can be obtained. Thus

$$\mu_3(V_n) = 24\mu^2\sigma^4/n^2 + O(n^{-5/2}), \qquad (2.10)$$

so that the standardized skewness of V_n is

$$\rho_3(V_n) = \mu_3(V_n)/\{\text{var}(V_n)\}^{3/2} = 3\sigma/(\mu\sqrt{n}) + O(n^{-1}). \qquad (2.11)$$

Similar expansions are obtained for W_n, although now the key expansion has error terms whose magnitude as $n \to \infty$ needs consideration. Thus, by Taylor's theorem with a remainder or by properties of the exponential function, it follows from (2.3) that

$$W_n - E(W_n) = e^{p\mu} p\sigma U_n^*/\sqrt{n} + R_n/n, \qquad (2.12)$$

where the remainder R_n has bounded moments. Thus

$$\{W_n - E(W_n)\}^2 = e^{2p\mu} p^2\sigma^2 U_n^{*2}/n + R_n'/n^{3/2},$$

where R_n' also has bounded moments. It now follows, on taking expectations, that

$$\text{var}(W_n) = e^{2p\mu} p^2\sigma^2/n + O(n^{-3/2}); \qquad (2.13)$$

if we want to calculate the term in $1/n^2$, exploration of R_n and R_n' is needed.

The present examples are atypical in that quite simple exact answers are available and the expansions therefore of relatively minor interest. It should be clear, however, that the techniques illustrated here are of wide potential applicability, in particular to cases where exact answers cannot readily be obtained. The essential points are that we start with a random variable U_n of exactly or approximately known distributional form and are interested in a function of U_n. Then we may be able by a local linearization to obtain a stochastic expansion for the new random variable, say in powers of $1/\sqrt{n}$ or $1/n$ and hence be able to deduce properties of the new random variable of interest from those of U_n.

2.3 Some convergence concepts

We now start a general development of the ideas exemplified in the previous section, the emphasis remaining, however, on concepts rather than on analytical detail. Let $\{U_n\}, \{V_n\}, \ldots$ be sequences of random variables, interest lying in large n; there is no necessity for n to represent a sample size.

We first review some definitions from elementary probability theory.

Convergence in probability

The sequence $\{U_n\}$ converges in probability to the constant a if given $\varepsilon > 0$ and $\eta > 0$, there exists $n_0 = n_0(\varepsilon, \eta)$ such that for any $n > n_0$

$$P\{|U_n - a| < \varepsilon\} > 1 - \eta. \tag{2.14}$$

That is, for sufficiently large n, almost all the probability distribution of U_n is concentrated very close to a. The associated notion in statistical theory is that of consistency of estimation. We often write $U_n \xrightarrow{\text{P}} a$.

A simple application of Tchebychev's inequality shows that if as $n \to \infty$

$$E(U_n) \to a, \qquad \text{var}(U_n) \to 0, \tag{2.15}$$

then U_n converges in probability to a. Note, however, that if $\text{var}(U_n) \not\to 0$, then, without further assumptions, no conclusion can be drawn from the behaviour of $E(U_n)$; for instance, we could have $E(U_n) \to b$, yet $U_n \xrightarrow{\text{P}} a \neq b$.

If $h(u)$ is a function continuous at $u = a$, it is easy to show that $U_n \xrightarrow{\text{P}} a$ implies that $h(U_n) \xrightarrow{\text{P}} h(a)$, the proof following directly from the definitions of continuity and convergence in probability. We have already had an instance of this result in section 2.2.

Convergence in distribution

The sequence $\{U_n\}$ is said to converge in distribution to the nondegenerate cumulative distribution function $F(u)$ if for all u that are continuity points of $F(u)$

$$\lim P(U_n \leqslant u) = F(u). \tag{2.16}$$

Equivalently one may require that $E\{h(U_n)\}$ tends to $E\{h(U)\}$ for all bounded and continuous functions $h(u)$, where U is a random variable having distribution function $F(u)$.

There are numerous differing notations in use. For example, one may write $\mathscr{L}(U_n) \to \mathscr{L}(U)$ and use the terminology 'convergence in law'. We shall use the notation

$$U \sim F, \quad U \sim W, \quad U_n \overset{\sim}{\to} F, \quad U_n \overset{\sim}{\to} U,$$

to mean respectively that U has distribution function F, that U and W have the same distribution function, that U_n converges in distribution to F and that U_n converges in law to the random variable U.

Note that restriction in (2.16) to continuity points of $F(u)$ is a minor technical detail arising because, in problems in which $F(u)$ is discrete, the value of $\lim P(U_n \leqslant u_0)$ is irrelevant when u_0 holds an atom of the limiting distribution $F(u)$. To understand this, consider the special case in which $P(U_n = 0) = \frac{1}{2}, P(U_n = 1 + \varepsilon_n) = \frac{1}{2}$, where $\{\varepsilon_n\}$ is a sequence with $\varepsilon_n \to 0$ and (a) $\varepsilon_n > 0$, (b) $\varepsilon_n < 0$.

Note also that convergence in law does not imply convergence of, or even existence of, moments. For example, if U_n has density

$$(1 - \varepsilon_n)(2\pi)^{-1/2} \exp(-\tfrac{1}{2}u^2) + \varepsilon_n \{\pi(1 + u^2)\}^{-1}, \qquad (2.17)$$

where, say, $\varepsilon_n = e^{-n}$, then convergence in law to N(0, 1) is extremely rapid, even though no moments exist. An important technical device for restoring the existence of moments and carrying through certain careful proofs is to truncate the distribution and to work with the moments and other properties of the truncated distribution. A detailed example concerning the truncated Poisson distribution is given later as Example 2.16.

Asymptotic normality

The sequence $\{V_n\}$ is asymptotically normal if there exist sequences of constants $\{a_n\}, \{b_n\}$ such that $(V_n - a_n)/b_n$ converges in distribution to the standard normal distribution. The constants a_n, b_n are called respectively the asymptotic mean and standard deviation of V_n.

We sometimes write $V_n \sim \mathrm{aN}(a_n, b_n^2)$. Asymptotic normality is concerned with the ability to approximate the quantiles of the distribution of V_n by those of $\mathrm{N}(a_n, b_n^2)$. In the earlier notation $(V_n - a_n)/b_n \xrightarrow{\mathcal{D}} \mathrm{N}(0, 1)$. As illustrated by (2.17) there need be no direct connexion between a_n, b_n and the mean and standard deviation of V_n, although of course in many well-behaved cases a_n and b_n could be taken as the mean and standard deviation of V_n.

While convergence to normality is the most commonly occurring case, very similar definitions and remarks apply for convergence to other distributions, of which the exponential, the chi-squared, the Poisson, the extreme value distributions and the stable laws are the most important special cases. In some cases standardization involves only a scale change, e.g. we may take $a_n = 0$ and choose b_n so that V_n/b_n has a limiting distribution.

Orders of magnitude

It is convenient to recall the o, O notation for sequences of constants $\{c_n\}, \{d_n\}$. We write $c_n = o(d_n)$ if $c_n/d_n \to 0$ as $n \to \infty$ and $c_n = O(d_n)$ if

c_n/d_n is bounded as $n \to \infty$. There is the useful convention that when the expression, say, $O(d_n)$ is used several times in an argument, different quantities may be involved on each occasion.

We may generalize the notation to random variables, writing $U_n = o_p(c_n)$ if $U_n/c_n \to 0$ in probability and $U_n = O_p(d_n)$ if U_n/d_n is bounded in probability. More explicitly, for the latter we require that given $\varepsilon > 0$ there exist constants k_ε and $n_0 = n_0(\varepsilon)$ such that if $n > n_0$

$$P\{|U_n| < d_n k_\varepsilon\} > 1 - \varepsilon.$$

In particular, $U_n = c + o_p(1)$ means that $U_n \xrightarrow{\text{p}} c$.

An important special case is when $\text{var}(U_n) \leqslant v/n$, for $n > n_0$, for some finite $v > 0$. Then $U_n = E(U_n) + O_p(n^{-1/2})$. If in addition $E(U_n) = \mu + O(n^{-1/2})$, then we have that

$$U_n = \mu + O_p(n^{-1/2}).$$

A result such as this specifies the rate of convergence in probability of U_n to μ.

The usual rules for manipulating o, O can be followed quite closely for o_p, O_p (see Further results and exercises 2.6).

Example 2.1 Sample moments

If $Y_1, \ldots, Y_n$ are independent and identically distributed random variables, copies of a random variable Y with $E(Y) = \mu$, $\text{var}(Y) = \sigma^2$, we have that

$$\begin{aligned} M_{n,2} &= \sum (Y_i - \bar{Y}_n)^2/n \\ &= \sum (Y_i - \mu)^2/n - (\bar{Y}_n - \mu)^2. \end{aligned}$$

Now $(\bar{Y}_n - \mu)^2 = O_p(1/n)$, so that

$$\sum (Y_i - \bar{Y}_n)^2/n = \sum (Y_i - \mu)^2/n + O_p(1/n). \tag{2.18}$$

Provided that the fourth moment of Y is finite, the central limit theorem applies directly to the second moment about μ, and when a standardized version of the random variable is introduced in the usual way, it is easy to see that the final term in (2.18) is negligible, so that the same limiting distribution applies to $M_{n,2}$ and also, of course, to the more usual estimate with divisor $n - 1$.

Example 2.2 Noncentral chi-squared

We define the noncentral chi-squared distribution with m degrees of freedom and noncentrality parameter Δ as the distribution of the random variable

$$W_m = (Z_1 + \Delta)^2 + Z_2^2 + \cdots + Z_m^2,$$

where $Z_1, \ldots, Z_m$ are independent and identically distributed random variables with the distribution N(0, 1). Consider its limiting behaviour as $\Delta \to \infty$ for fixed m. Incidentally, one of the mathematical interests of the noncentral chi-squared distribution is that a number of other limiting operations can be considered.

Now

$$W_m = \Delta^2 + 2\Delta Z_1 + Z_1^2 + \cdots + Z_m^2.$$

We standardize by writing

$$W_m^* = (W_m - \Delta^2)/(2\Delta). \tag{2.19}$$

It follows that

$$W_m^* = Z_1 + O_p(1/\Delta), \tag{2.20}$$

so that as $\Delta \to \infty$ not only do we have $W_m^* \xrightarrow{d} \mathrm{N}(0, 1)$, but W_m is in the limit essentially determined by Z_1.

2.4 Convergence after functional transformation

Before discussing generalities on asymptotic expansions, we state an important argument connected with convergence in distribution.

First suppose that U_n converges in distribution to the nondegenerate limiting distribution function $F(u)$ and that $V_n \xrightarrow{\mathrm{p}} c$. Then in distribution

$$U_n + V_n \xrightarrow{d} F(u - c). \tag{2.21}$$

If U has distribution function $F(u)$, we can write $U_n + V_n \xrightarrow{d} U + c$.

The right-hand side of (2.21) is the limiting distribution of $U_n + c$. Note that $U_n = O_p(1)$ and $V_n = c + o_p(1)$. Similarly, although the proof is slightly more complicated, if $c > 0$,

$$U_n V_n \xrightarrow{d} Uc, \qquad U_n/V_n \xrightarrow{d} U/c, \tag{2.22}$$

the corresponding distribution functions being $F(u/c)$ and $F(uc)$.

These results are subsumed in the following theorem. We recall that the support of a random variable U is the set of all u such that $P(u - \varepsilon < U < u + \varepsilon) > 0$ for all $\varepsilon > 0$.

Theorem

Let $U_n \xrightarrow{\mathcal{D}} U$, having distribution function $F(u)$, and let V_n tend in probability to c. Let $h(U_n, V_n)$ be such that $h(u, v)$ is a continuous function of v at $v = c$ for all points u in the support of U. Then $h(U_n, V_n) \xrightarrow{\mathcal{D}} h(U, c)$.

A multivariate version of this theorem is discussed in section 2.9.

Example 2.3 Student's t

Let $Y_1, \ldots, Y_n$ be independent and identically distributed random variables, copies of a random variable Y with $E(Y) = \mu$, $\text{var}(Y) = \sigma^2$ and with finite fourth moment. Let $\bar{Y}_n = \sum Y_j/n$, $s_n^2 = \sum(Y_j - \bar{Y}_n)^2/(n-1)$ and consider the Student t statistic defined by

$$T_n = \sqrt{n}(\bar{Y}_n - \mu)/s_n. \tag{2.23}$$

Now $E(s_n^2) = \sigma^2$, $\text{var}(s_n^2) \to 0$, so that $s_n^2 \xrightarrow{\text{P}} \sigma^2$ and hence $s_n \xrightarrow{\text{P}} \sigma$. Further by the central limit theorem,

$$U_n = \sqrt{n}(\bar{Y}_n - \mu)$$

is asymptotically $\text{N}(0, \sigma^2)$. Thus if in the theorem we take $V_n = s_n$, we may conclude that T_n is asymptotically $\text{N}(0, 1)$.

The finiteness of μ_4 enters only in proving that $\text{var}(s_n^2) \to 0$; in fact a more elaborate argument shows that convergence follows from the finiteness of σ^2. The result leaves open the questions of speed of convergence to the limiting distribution and of the appropriateness of the statistic T_n in the analysis of nonnormal data.

2.5 Approximate linearization

In the previous section, we outlined a key argument in which convergence in probability was combined with convergence in distribution to isolate the limiting behaviour of a function of two or more random variables. A second key element in the discussion hinges on the local linearization of functions, often referred to as the δ-method.

Suppose that $V_n \xrightarrow{\text{P}} a$ and that $\sqrt{n}(V_n - a) \xrightarrow{\mathcal{D}} F$. Thus we can write

$$\sqrt{n}(V_n - a) \sim V^* + o_p(1),$$

i.e.

$$V_n \sim a + V^*/\sqrt{n} + o_p(1/\sqrt{n}), \tag{2.24}$$

where V^* has distribution function $F(v)$. In many applications F is the standard normal distribution; also, if n is sample size, the sequence $\sqrt{n}$ may sometimes have to be replaced by some other normalizing sequence. For example, if the regression coefficient (1.10) is of interest, the use of $n^{3/2}$ will be called for under the special assumptions of Example 1.1. Note also that there is a technical issue here concerning the space on which the new random variable V^* is defined. The essential point is that we can define V^* so that the two sides of (2.24) have the same distribution.

Let $h(v)$ be a twice differentiable function with $h'(a) \neq 0$. Initially we for simplicity shall impose the unnecessarily strong regularity requirement of a uniformly bounded second derivative on the support of $\{V_n\}$ for $n > n_0$. Consider the sequence $\{h(V_n)\}$.

Now consider

$$H_n^* = \sqrt{n}\{h(V_n) - h(a)\} \\ = \sqrt{n}h'(a)(V_n - a) + R_n, \tag{2.25}$$

where

$$R_n = \tfrac{1}{2}\sqrt{n}h''\{\xi a + (1 - \xi)V_n\}(V_n - a)^2 \tag{2.26}$$

with $0 < \xi < 1$. That is, the remainder is determined by the second derivative at a point intermediate between a and V_n. The first term of (2.25) is

$$V^*h'(a) + o_p(1)$$

and the second term, via (2.24) and the boundedness of h'' is $O_p(n^{-1/2})$. That is,

$$H_n^* = \sqrt{n}\{h(V_n) - h(a)\} = V^*h'(a) + o_p(1) \tag{2.27}$$

so that $H_n^* \stackrel{\mathcal{D}}{\to} V^*h'(a)$. In particular, if V_n is asymptotically normal so is $h(V_n)$.

Example 2.4 A squared random variable

Let $h(u) = u^2$, so that h'' is indeed bounded. Then if V_n is $\mathrm{aN}(\mu, \sigma^2/n)$ with $\mu \neq 0$, then V_n^2 is $\mathrm{aN}(\mu^2, 4\mu^2\sigma^2/n)$.

For simplicity of exposition, we have given the above argument with the expansion (2.25) applying for all values of V_n ($n > n_0$) and with a uniformly bounded h''. It is clear, however, that these conditions can be relaxed. Thus if we define R_n by (2.26) we have only to prove that $R_n \xrightarrow{p} 0$.

Example 2.5 Square root

Let $h(v) = \sqrt{v}$. Then, if V_n is aN$(\mu, \kappa/n)$ with $\mu > 0$,

$$8R_n = -\sqrt{n}\{\xi\mu + (1-\xi)V_n\}^{-3/2}(V_n - \mu)^2.$$

Because $n(V_n - \mu)^2$ converges in distribution, it follows that $R_n \xrightarrow{p} 0$ and therefore V_n is aN$\{\mu, \kappa/(4\mu n)\}$. For example, if V_n is the usual estimate of variance from n independent and identically distributed random variables from a distribution with finite fourth cumulant, it follows from the central limit theorem that V_n is aN$\{\sigma^2, 2\sigma^4(1 + \frac{1}{2}\rho_4)/n\}$. Therefore $\sqrt{V_n}$, the estimate of standard deviation, is aN$\{\sigma, \frac{1}{2}\sigma^2(1 + \frac{1}{2}\rho_4)/n\}$.

Example 2.6 Variance stabilization

Let X_n be a random variable with a distribution within a family such that $E(X_n) = \theta$, $\text{var}(X_n) = v_n(\theta) = v(\theta)/n$, where $v(\theta)$ is a known function and n defines a notional sequence of problems. For example, if X_n is the number of successes in n independent Bernoulli trials with probability of success θ, $v(\theta) = \theta(1-\theta)$. In other instances $v(\theta)$ may be determined empirically. We now look for a function $h(x)$ such that $h(X_n)$ has asymptotically a variance, say $1/n$, not depending on θ.

The asymptotic variance of $h(X_n)$ is $\{h'(\theta)\}^2 v(\theta)/n$, so that we require

$$h'(\theta) = 1/\sqrt{v(\theta)},$$

$$h(\theta) = \int_a^\theta \{1/\sqrt{v(t)}\}\, dt, \tag{2.28}$$

for some suitable a. The most familiar example of this is when $v(x) = k^2 x^{2m}$, leading in particular when $m = 1$ to

$$h(x) = \log x. \tag{2.29}$$

Example 2.7 Maximum likelihood estimate (scalar parameter)

A further illustration of these formulae is provided by the theory of regular maximum likelihood estimation. Let $Y_1, \ldots, Y_n$ be independent and identically distributed random variables with density $f(y;\theta)$. Then it is known that under suitable regularity conditions the maximum likelihood estimate $\hat{\theta}$ is $\text{aN}\{\theta, i_\theta^{-1}(\theta)/n\}$, where

$$i_\theta(\theta) = E\{-\partial^2 \log f(Y;\theta)/\partial\theta^2\}$$
$$= E\{(\partial \log f(Y;\theta)/\partial\theta)^2\}.$$

Now suppose that ϕ is a differentiable strictly monotonic function of θ. Then $\hat{\phi} = \phi(\hat{\theta})$ and it follows that $\hat{\phi}$ is asymptotically normal with mean $\phi(\theta)$ and variance

$$(d\phi/d\theta)^2 i_\theta^{-1}(\theta)/n. \tag{2.30}$$

But if we define the information $i_\phi(\phi)$ for the new parameterization in the standard way, we have that

$$i_\phi(\phi) = E\{(\partial \log f(Y;\phi)/\partial\phi)^2\}$$
$$= (d\theta/d\phi)^2 i_\theta(\theta),$$

showing that the asymptotic variance of $\hat{\phi}$ is indeed

$$i_\phi^{-1}(\phi)/n,$$

establishing the 'self-consistency' of (2.30).

More elaborate discussions are necessary to establish how the speed of convergence to asymptotic normality depends on the particular parameterization.

There is a very close formal parallel with the discussion of variance stabilization in Example 2.6 obtained by finding a parameterization ϕ for which the information is constant. For this we require

$$d\phi/d\theta = \sqrt{i_\theta},$$

i.e.

$$\phi = \int_a^\theta \sqrt{i_\theta(\lambda)}d\lambda.$$

Now for the one-parameter exponential family model (1.15), $i(\theta) = k''(\theta)$, from which the information-stable parameterization follows immediately for many standard distributions. For instance, for the Poisson distribution the square root of the mean is the parameter indicated.

The above results all depend on the condition $h'(a) \neq 0$, so that there is a nontrivial linear approximation to $h(x)$ in the neighbourhood under study. If $h'(a) = 0$, the Taylor expansions have to be taken to further terms and the form of the asymptotic behaviour is essentially settled by the order of the first nonvanishing derivative.

Example 2.8 A trigonometric function

Let $\{V_n\}$ be $\mathrm{aN}(0, \sigma^2/n)$ and let $h(v) = \cos v$. Then $V_n = \sigma V^*/\sqrt{n} + o_p(1/\sqrt{n})$, where V^* has the standard normal distribution. Now

$$\cos V_n = 1 - \tfrac{1}{2}\sigma^2 V^{*2}/n + o_p(n^{-1}),$$

showing that $2n(1 - \cos V_n)/\sigma^2$ converges to the chi-squared distribution with one degree of freedom. Note particularly the occurrence of n rather than $\sqrt{n}$ as a normalizing constant.

Example 2.9 A squared random variable

An even simpler example is provided by taking $h(v) = v^2$, with $\{V_n\}$ being $\mathrm{aN}(0, \sigma^2/n)$. The linearization around the mean μ of V_n no longer works when V_n itself has zero mean. We have that $\sqrt{n}V_n/\sigma = V_n^*$, where $V_n^* \xrightarrow{\mathscr{D}} \mathrm{N}(0, 1)$, so that nV_n^2 is asymptotically proportional to chi-squared with one degree of freedom.

Comparison with the earlier discussion in Example 2.2 of $h(v) = v^2$ reveals a discontinuity of behaviour. If $\{V_n\}$ is $\mathrm{aN}(\mu, \sigma^2/n)$, then if $\mu = 0$ not only is a different standardization by n rather than by $\sqrt{n}$ required but the limiting distribution is quite different.

This raises a difficulty if the limiting distribution is to be used as an approximation for a particular n when V_n has 'small' mean and 'small' standard deviation and it is natural to look in such cases for an intermediate regime to bridge the transition between the limiting forms. It is difficult to give a formal procedure by which this is to be done, although in particular cases it will often be clear how to proceed. In the present instance we suppose that for some γ, $V_n - \gamma/\sqrt{n}$ is $\mathrm{aN}(0, \sigma^2/n)$. Note that the fluctuations in V_n are $O_p(1/\sqrt{n})$, so that the term $\gamma/\sqrt{n}$ represents a displacement of magnitude comparable with a typical fluctuation. Thus if $V^* \sim \mathrm{N}(0, 1)$,

$$V_n = \gamma/\sqrt{n} + (\sigma/\sqrt{n})V^* + o_p(1/\sqrt{n}),$$

i.e.

$$nV_n^2 = \sigma^2(V^* + \gamma/\sigma)^2 + o_p(1),$$

so that nV_n^2/σ^2 converges to noncentral chi-squared with one degree of freedom. Thus if $\gamma = 0$, we recover the central chi-squared limit. If we start with the noncentral chi-squared distribution and let the noncentrality parameter tend to infinity, writing $\Delta^2 = n\mu^2/\sigma^2$ and letting $n \to \infty$ for fixed μ, we recover after a further standardization a normal limit which is the result of Example 2.2.

It is an indication of both the power and the limitations of the results that if, say, $\{V_n\}$ is aN$(\mu, \sigma^2/n)$, then any reasonable function of V_n also is asymptotically normal, provided only that the slope is nonzero at μ. For instance, if s_n^2 is the usual estimate of variance from independent and identically distributed random variables, not only is $\{s_n^2\}$ asymptotically normal, but so too are $\{s_n\}$, $\{\log s_n\}$, $\{s_n^{2/3}\}, \ldots$, because of the local linearity. This is a powerful conclusion: on the other hand, it emphasizes too the need both for the examination of higher-order approximations and for numerical work if conclusions are to be drawn about relative closeness to normality of the different functions.

2.6 Direct calculations with density functions and moment generating functions

There are broadly three ways of examining the asymptotic properties of distributions, directly from the form of the random variable, by analysis of the density or cumulative distribution function and by study of the moment or cumulant generating function. The last is obviously particularly convenient when the generating function is available in explicit form, but rests on deep analytical results establishing the expansions or limiting results for distributions.

The most familiar example of that argument concerns the central limit theorem for the sum of independent and identically distributed random variables, when, after standardization, elementary expansion gives the limiting cumulant generating function as $\frac{1}{2}t^2$, that of the standard normal distribution. We shall not give the details here, partly because we shall be considering a generalization in Chapter 4. We give instead some other relatively elementary examples.

Example 2.10 Gamma distribution of large index

We examine the density

$$f_n(x) = \lambda(\lambda x)^{n-1} e^{-\lambda x}/(n-1)! \tag{2.31}$$

as $n \to \infty$ for fixed λ.

Note first that (2.31) is the density of the sum S_n of n independent and identically distributed random variables each with the density $\lambda e^{-\lambda x}$, with mean and standard deviation $1/\lambda$. Therefore, by the central limit theorem, (2.31) tends to the normal density of mean n/λ and variance n/λ^2; more strictly, of course, we have either to use a so-called local version of the central limit theorem applying directly to the density, or to phrase the conclusion in terms of the cumulative distribution function of the standardized random variable.

A second argument works directly with the explicit form of the density (2.31). We introduce the standardizing transformation

$$x = n/\lambda + z\sqrt{n}/\lambda \qquad (2.32)$$

and then express $\log f_n(x)$ as a function of the new argument z in the form

$$\log f_n(x) = n \log \lambda + (n-1)\log(n/\lambda + z\sqrt{n}/\lambda) - n - z\sqrt{(n\lambda)} - \log(n-1)!. \qquad (2.33)$$

Elementary, if mildly tedious, expansion shows that

$$\log f_n(x) = -\tfrac{1}{2}\log(2\pi n/\lambda) - \tfrac{1}{2}z^2 + o(1) \qquad (2.34)$$

which corresponds to the limiting normal density of the standardized random variable, the factor n/λ being removed when we transform to that random variable.

A third possibility is to work with the cumulant generating function of S_n, to transform to the standardized random variable and then to expand; this in effect reproduces in a special case the standard argument for the central limit theorem.

Example 2.11 Extreme value distribution

An instance where direct calculation with the density or distribution function is needed is provided by the limiting distribution of extreme order statistics. Here we discuss only briefly just one special case. Suppose that $X_1, X_2, \ldots$ are independent and identically distributed random variables, copies of a random variable X. Suppose that the support of X has a lower terminal, say at zero, and that as $x \to 0$ the cumulative distribution function of X is such that

$$F(x) \sim ax^b \qquad (a, b > 0). \qquad (2.35)$$

Now let $Y_n = \min(X_1, \ldots, X_n)$ and consider the limiting distribution of Y_n as $n \to \infty$.

It is clear that with high probability Y_n takes only small values, so that standardization is needed to achieve a useful limiting result. Standardization by mean and standard deviation is not available and is in any case inappropriate; only local behaviour near 0 is relevant. Therefore we try $Y_n^* = n^c Y_n$, aiming to choose c suitably.

Now

$$P(Y_n \leqslant y) = 1 - P(X_1 > y, \ldots, X_n > y)$$
$$= 1 - \{1 - F(y)\}^n,$$

so that

$$P(Y_n^* \leqslant y) = P(Y_n \leqslant y/n^c)$$
$$= 1 - \{1 - F(y/n^c)\}^n$$
$$\sim 1 - (1 - ay^b/n^{bc})^n.$$

Thus we choose $bc = 1$, leading to

$$P(Y_n^* \leqslant y) \to 1 - \exp(-ay^b), \tag{2.36}$$

showing that the standardized form $n^{1/b} \min(X_1, \ldots, X_n)$ has a limiting Weibull distribution.

There is an extensive literature on the precise conditions for this to hold and on the corresponding conditions for the other two possible limit laws for extremes.

2.7 Multidimensional version

While in this chapter we concentrate on univariate problems, it is useful to record that many of the arguments extend immediately to finite-dimensional random variables. In particular, $\{U_n\}$ is a sequence of m-dimensional random variables converging in distribution with limiting distribution function $F(u)$ if at all continuity points of F

$$\lim P(U_{n1} \leqslant u_1, \ldots, U_{nm} \leqslant u_m) = F(u_1, \ldots, u_m),$$

where $U_{n1}, \ldots, U_{nm}$ denote the components of U_n. Often F will be $\mathrm{MN}_m(\mu, \Omega)$, the multivariate normal distribution in m dimensions of mean μ and covariance matrix Ω. That is, the ith component of μ and the (i, j)th component of Ω give respectively the mean of the ith random variable and the covariance of the ith and jth random variables.

Now suppose that the sequence of random vectors $\{V_n\}$ is such that $\sqrt{n}(V_n - a)$ converges in distribution to a multivariate normal distribution of zero mean and covariance matrix Ω, $\mathrm{MN}_m(0, \Omega)$, and that $h(v)$ is a real-valued function of v with nonvanishing gradient at a. Then under assumptions about second derivatives similar to those in the one-dimensional case,

$$\begin{aligned} H_n^* &= \sqrt{n}\{h(V_n) - h(a)\} \\ &= \sqrt{n}(\nabla h)^{\mathrm{T}}(V_n - a) + R_n, \end{aligned} \tag{2.37}$$

where $\nabla h = \nabla h(a)$ is the $m \times 1$ gradient of h at $v = a$, i.e.

$$\nabla h(a) = [\partial h(v)/\partial v_1, \ldots, \partial h(v)/\partial v_m]^{\mathrm{T}}_{v=a}.$$

Provided that $R_n = o_p(1)$, we have that

$$H_n^* = (\nabla h)^{\mathrm{T}} V^* + o_p(1), \tag{2.38}$$

where V^* has the distribution $\mathrm{MN}_m(0, \Omega)$. Thus H_n^* converges to a normal distribution of zero mean and variance $(\nabla h)^{\mathrm{T}}\Omega\nabla h$, i.e.

$$h(V_n) \sim \mathrm{aN}\{h(a), (\nabla h)^{\mathrm{T}}\Omega\nabla h/n\}. \tag{2.39}$$

In a more general formulation of these ideas, we have q functions $h_1, \ldots, h_q$ of the m real variables $U_1, \ldots, U_m$, i.e. h is a function (mapping) from R^m to R^q; equation (2.39) still applies and gives the asymptotic mean and covariance matrix.

For further discussion of the mathematical aspects, see section 2.9.

Example 2.12 Ratio of random variables

Suppose that $V_n = (V_{n1}, V_{n2})$ is such that $\sqrt{n}(V_n - \mu)$ is asymptotically bivariate normal with zero mean and covariance matrix Ω, where $\mu = (\mu_1, \mu_2)$ with $\mu_2 \neq 0$. Let $h(V_n) = V_{n1}/V_{n2}$. Now $\nabla h(\mu) = (1/\mu_2, -\mu_1/\mu_2^2)^{\mathrm{T}}$. Then $h(V_n)$ is asymptotically normal with mean μ_1/μ_2 and variance

$$(1/\mu_2, -\mu_1/\mu_2^2)\Omega(1/\mu_2, -\mu_1/\mu_2^2)^{\mathrm{T}} n^{-1}. \tag{2.40}$$

Note especially that if and only if $\mu_1 = 0$, (2.40) depends just on ω_{11}, i.e. the asymptotic distribution is exactly that of V_{n1}/μ_2, in accord with the results of section 2.4.

There are many instances in statistical inference where an asymptotically normal random variable of zero mean is divided by an estimate of its standard error and the permissible neglect of errors of estimation of the standard error is the basis of many widely used simple formulae.

Example 2.13 Product of random variables

Suppose now that in the discussion of the previous example we replace the ratio of components by the product, i.e. $k(V_n) = V_{n1} V_{n2}$. Then $\nabla k = (\mu_2, \mu_1)^{\mathrm{T}}$ and $V_{n1} V_{n2}$ is

$$\mathrm{aN}\{\mu_1 \mu_2, (\mu_2^2 \omega_{11} + 2\mu_1 \mu_2 \omega_{12} + \mu_1^2 \omega_{22})/n\} \tag{2.41}$$

provided that the asymptotic variance as stated is nonzero and in particular that $(\mu_1, \mu_2) \neq (0, 0)$. Now in this special case moments of the product can be calculated exactly from the mixed moments of the components, so that some immediate check on the adequacy of the asymptotic result is available. In particular, in the special case where μ and Ω/n are the exact mean and covariance matrix of V_n and the distribution is exactly bivariate normal, then

$$E(V_{n1} V_{n2}) = \mu_1 \mu_2 + \omega_{12}/n,$$

$$n \operatorname{var}(V_{n1} V_{n2}) = \omega_{11} \mu_2^2 + 2\omega_{12} \mu_1 \mu_2 + \omega_{22} \mu_1^2 + (\omega_{11} \omega_{22} + \omega_{12}^2)/n.$$

Example 2.14 Joint distribution of ratio and product

Suppose now that we consider the joint distribution of $(V_{n1}/V_{n2}, V_{n1} V_{n2})$. With $\mu_2 \neq 0$, we have asymptotic bivariate normality and it remains only to find the covariance in the limiting distribution. By (2.39) it is

$$(1/\mu_2, -\mu_1/\mu_2^2)^{\mathrm{T}} \Omega(\mu_2, \mu_1) = (\omega_{11} - \omega_{22} \mu_1^2/\mu_2^2) n^{-1}. \tag{2.42}$$

Thus a necessary and sufficient condition for the asymptotic independence of ratio and product is that $\omega_{11}/\mu_1^2 = \omega_{22}/\mu_2^2$, i.e. that the squared coefficients of variation of the two components are asymptotically equal. For positive random variables, an alternative simpler proof follows by examining $\log V_{n1} - \log V_{n2}$ and $\log V_{n1} + \log V_{n2}$.

Example 2.15 Transformation of covariance matrices

In Example 2.6 we studied transformation of a univariate random variable to approximately constant variance when the variance of the original random variable is a known function of the mean. A natural multivariate generalization of this problem involves an $m \times 1$ random variable X_n such that $E(X_n) = \mu$, $\operatorname{cov}(X_n) = \Omega(\mu)/n$, where $\Omega(\mu)$ is a known function. We now look for a transformation from

X_n to $h(X_n)$, a $1 \times m$ row vector of functions, such that cov $\{h(X_n)\}$ is asymptotically a constant matrix, which without loss of generality we may take to be the identity matrix.

For this to be achieved we must have

$$(\nabla h)^{\mathrm{T}}\Omega(\mu)(\nabla h) = I, \tag{2.43}$$

where the derivative matrix is evaluated at μ. Unfortunately these equations are in general inconsistent, it being impossible to satisfy the compatibility equations exemplified by

$$\partial/\partial x_1(\partial h/\partial x_2) = \partial/\partial x_2(\partial h/\partial x_1).$$

A geometric interpretation of this is obtained by regarding Ω as defining the metric tensor of a Riemannian geometry; it is only when the space in question is Euclidean that the required transformation exists. Moolgavkar and Venzon (1987) have, however, pointed out that an approximate solution is available via use of geodesic normal coordinates in an appropriate geometry.

2.8 Convergence of moments

In the previous discussion of local linearization the emphasis is on convergence in distribution. For example, we are concerned with approximations valid for fixed x and large n to

$$P[\sqrt{n}\{h(V_n) - h(a)\} \leqslant x] = P\{h(V_n) \leqslant h(a) + x/\sqrt{n}\}. \tag{2.44}$$

This is, of course, not the same as deriving approximations for $E\{h(V_n)\}$, var $\{h(V_n)\}$, etc., which may be seriously influenced by behaviour in the extreme tail of the distribution. On the whole, however, at least in statistical applications, it is approximations to distributions that are required, rather than those to moments. This is seen most clearly in those cases where, say, a test statistic has to very close approximation a simple 'well-behaved' distribution yet divergent mean or variance. An example connected with normal theory linear regression arises in estimating a value of the explanatory variable given a value of the response variable, the problem of inverse regression. Because of the presence of the reciprocal of an estimated regression coefficient in the natural point estimate, one is led to consider a random variable of infinite variance and yet answers based on a normal approximation may be entirely adequate.

If in the linearizing approximation (2.25) the remainder term has moments that behave suitably, as is probably the case in the great majority of specific practical applications, approximations to the moments can be obtained by direct evaluation or, sometimes, by manipulation of the cumulant generating function. Even when the conditions fail we can usually regard the expansions as holding for a suitably truncated random variable.

We shall give just two simple examples.

Example 2.16 Reciprocal of Poisson variable

Suppose that V_n has a Poisson distribution of mean n and that $W_n = h(V_n) = 1/V_n$.

Formal application of the expansion would suggest that

$$E(W_n) \doteq 1/n, \tag{2.45}$$

whereas in fact W_n is an improper random variable for all n, because $P(V_n = 0) > 0$.

On the other hand, use of a normal approximation to find the distribution function of W_n gives good results even for quite small values of n. Also if we truncate V_n by omitting the zero value, i.e. consider a random variable V'_n such that

$$P(V'_n = r) = e^{-n}n^r/\{(1-e^{-n})r!\} \qquad (r = 1, 2, \ldots) \tag{2.46}$$

then all is in order. Grab and Savage (1954) have tabulated $E(1/V'_n)$; the values at $n = 5, 10, 20$ are respectively 0.258, 0.113, 0.053, illustrating the increasingly good approximation by $1/n$.

Example 2.17 Functions of normal theory estimates of variance

A normal theory estimate of variance, T_m, with m degrees of freedom is, except for a scale constant, m^{-1} times a chi-squared random variable, W_m, with m degrees of freedom. Since $E(W_m) = m$, $\text{var}(W_m) = 2m$, a standardized version of T_m and W_m is

$$V^*_m = T^*_m = (W_m - m)(2m)^{-1/2}. \tag{2.47}$$

Now suppose that we are interested in the corresponding estimate of standard deviation, $\sqrt{T_m}$. It is easily shown that

$$E(\sqrt{T_m}) = \{\Gamma(\tfrac{1}{2}m + \tfrac{1}{2})/\Gamma(\tfrac{1}{2}m)\}(2/m)^{1/2}. \tag{2.48}$$

If, for example, we were interested in the bias of the estimate of standard deviation we could thus proceed by numerical evaluation of (2.48), or by applying Stirling's formula (see Example 3.5) for approximating the gamma functions. A method essentially equivalent to the last but of broader applicability is to work directly with an expansion for the moments as follows. We have in terms of the standardized form of T_m or W_m that

$$\begin{aligned} E(\surd T_m) &= E\{(1 + W_m^* \surd(2/m)\}^{1/2} \\ &= E\{1 + \tfrac{1}{2} W_m^* \surd(2/m) - \tfrac{1}{8} W_m^{*2}(2/m) + \cdots\} \\ &= 1 - 1/(4m) + \cdots. \end{aligned} \tag{2.49}$$

Further terms can be obtained but care is needed to ensure that all the contributions of a given order in $1/m$ are retained; it is for this reason that the explicit introduction of the standardized form is helpful.

Another important example is provided by $\log T_m$.

2.9 Appendix: Some basic limit theorems

We provide here concise and general versions of the basic limit theorems discussed in the previous part of this chapter.

We begin by restating formally the definitions of convergence in distribution or weak convergence and of convergence in probability. Let X and X_n $(n = 1, 2, \ldots)$ be m-dimensional random variables and let Q and Q_n be the corresponding probability measures on R^m. Furthermore, let $\bar{C}(R^m)$ denote the set of bounded and continuous real valued functions on R^m.

Definition 2.1 Convergence in distribution

The sequence of random vectors X_n converges in distribution (or converges weakly) to X, in symbols $X_n \xrightarrow{d} X$, if

$$\int f\,dQ_n \rightarrow \int f\,dQ \quad \text{for all } f \in \bar{C}(R^m),$$

i.e. $E\{f(X_n)\} \rightarrow E\{f(X)\}$.

Let ϕ and ϕ_n denote the characteristic functions of X and X_n, respectively, e.g. $\phi(t) = E(e^{itX})$.

Theorem 2.1

If $X_n \xrightarrow{d} X$ then $\phi_n(t) \rightarrow \phi(t)$ uniformly on compact subsets of R^m. On the other hand, if $\phi_n(t) \rightarrow \phi(t)$ for all $t \in R^m$ then $X_n \xrightarrow{d} X$.

For a proof, see Renyi (1970; sections 6.4 and 6.6). Note that this assumes the existence of X. If it is required to deduce the existence of a proper limiting distribution from the convergence of the characteristic functions, continuity of $\phi(t)$ at $t=0$ is needed.

Definition 2.2 Convergence in probability
The sequence of random vectors X_n converges in probability to X, in symbols $X_n \xrightarrow{p} X$, if

$$P\{|X_n - X| > \varepsilon\} \to 0 \quad \text{as } n \to \infty, \quad \text{for all } \varepsilon > 0,$$

where $|.|$ denotes Euclidean norm.

By far the most important special case is when $X_n \xrightarrow{p} a$, where a is a constant; see section 2.3. The more general notion requires that X is defined on the same probability space as the sequence $\{X_n\}$. A natural example arises if $\{Y(t)\}$ is a stochastic process in continuous time; then, for well-behaved processes, for all t, $Y(t + c/n) \xrightarrow{p} Y(t)$ as $n \to \infty$ for fixed c.

Theorem 2.2
If $X_n \xrightarrow{p} X$ then $X_n \xrightarrow{\mathcal{D}} X$. As a partial converse we have that if X is constant, $X = c$, then $X_n \xrightarrow{\mathcal{D}} c$ implies $X_n \xrightarrow{p} c$.

Now, let $U_n = (U_{n1}, \ldots, U_{nq})$ and $V_n = (V_{n1}, \ldots, V_{nr})$ be two sequences of random vectors, and let h denote a (measurable) mapping from one Euclidean space R^q into another R^s, i.e. h denotes a vector of s (not bizarre) real-valued functions of q real variables $(u_1, \ldots, u_q)$. Further, let c denote a constant vector.

Since convergence in probability to a constant is a special case of convergence in law, the contents of the first three theorems below can be summarized as saying that a continuous (vector-valued) function of a (vector-valued) random variable converging in law itself converges in law. As the proofs are quite similar, we verify only Theorem 2.5.

Theorem 2.3
Suppose $h: R^q \to R^s$ is continuous at $c \in R^q$ and $U_n \xrightarrow{p} c$. Then $h(U_n) \xrightarrow{p} c$.

Theorem 2.4
Suppose $h: R^q \to R^s$ is continuous and $U_n \xrightarrow{\mathcal{D}} U$. Then $h(U_n) \xrightarrow{\mathcal{D}} h(U)$.

Theorem 2.5 (Continuous transformation theorem)
Suppose $h: R^{q+r} \to R^s$ is continuous, and $U_n \rightsquigarrow U, V_n \rightsquigarrow c$. Then $h(U_n, V_n) \rightsquigarrow h(U, c)$. In particular, $(U_n, V_n) \rightsquigarrow (U, c)$.

Outline of proof. Let P_n, P and Q_n, Q be the probability measures on R^{q+r} and R^s determining the distributions of (U_n, V_n), (U, c) and $h(U_n, c)$, $h(U, c)$, respectively, and let $f \in \bar{C}(R^s)$.

It suffices to show that $(U_n, V_n) \rightsquigarrow (U, c)$, for having established this the general result follows on noting that the composed function $g = f \circ h$ belongs to $\bar{C}(R^{q+r})$, whence

$$\int f dQ_n = \int g\, dP_n \to \int g\, dP = \int f dQ.$$

It is simple to see that $(U_n, c) \rightsquigarrow (U, c)$ and since for any $g \in \bar{C}(R^{q+r})$

$$g(U_n, V_n) = \{g(U_n, V_n) - g(U_n, c)\} + g(U_n, c)$$

the result will follow by proving that

$$g(U_n, V_n) - g(U_n, c) \xrightarrow{\mathrm{p}} 0.$$

Using the fact that g is uniformly continuous on compact sets the latter convergence is not difficult to establish.

Recall that a mapping $h: R^q \to R^s$, defined in a neighbourhood of $u_0 \in R^q$, is said to be totally differentiable a′. u_0 provided there exists a $s \times q$ matrix $J(u_0)$ such that

$$h(u) = h(u_0) + J(u_0)(u - u_0) + o(|u - u_0|), \qquad (2.50)$$

where $|\cdot|$ indicates ordinary Euclidean distance. In this case the partial derivatives of h exist at u_0 and

$$J(u_0) = \nabla h(u_0), \qquad (2.51)$$

the $s \times q$ gradient matrix of h at u_0 consisting of the first-order partial derivatives of h. Conversely, if the partial derivatives exist and are continuous in a neighbourhood of u_0 then h is totally differentiable at u_0.

Theorem 2.6 (Differentiable transformation theorem)
Suppose $h: R^q \to R^s$ is totally differentiable at $u_0 \in R^q$ and that $\nabla h(u_0)$ has rank s. If $\sqrt{n}(U_n - u_0) \rightsquigarrow \mathrm{MN}_q(0, \Omega)$ then

$$\sqrt{n}\{h(U_n) - h(u_0)\} \rightsquigarrow \mathrm{MN}_s(0, \nabla h(u_0) \Omega \nabla h(u_0)^{\mathrm{T}}).$$

Outline of proof. We have $o_p(\sqrt{n}|U_n - u_0|) \xrightarrow{p} 0$ and hence the result follows from Theorem 2.5 together with (2.50) and (2.51) if we show that

$$\sqrt{n}J(u_0)(U_n - u_0) \rightsquigarrow \text{MN}_s(0, J(u_0)\Omega J(u_0)^{\text{T}}),$$

and this is an immediate consequence of Theorem 2.1.

A key point of Theorem 2.6 is that differentiability alone is enough. Of course, rate of approach to the limiting result is determined by further features.

Further results and exercises

2.1 Prove that if $E(U_n) \to a$, $\text{var}(U_n) \to 0$, then U_n tends in probability to a. Give simple examples to show that neither condition is necessary for the conclusion.

Prove that if $U_n \to a$ in probability and $h(u)$ is continuous at $u = a$, then $h(U_n) \to h(a)$ in probability. What happens if $h(u)$ is discontinuous at $u = a$ but has separate left- and right-hand limits?

[Sections 2.2, 2.3]

2.2 Formulate explicitly and prove the properties that are needed to justify the statement in Example 2.1 that $\sum(Y_i - \mu)^2/n$, $M_{n,2} = \sum(Y_i - \bar{Y}_n)^2/n$ and $\sum(Y_i - \bar{Y}_n)^2/(n-1)$ have the same asymptotic distribution.

[Section 2.3]

2.3 The simplest conditions for the asymptotic normality of the sum $S_n = Y_1 + \cdots + Y_n$ of independent random variables are

(i) if the $\{Y_j\}$ are identically distributed copies of a random variable Y, it is sufficient that $\text{var}(Y) < \infty$;
(ii) if the $\{Y_j\}$ are not identically distributed, it is sufficient that the absolute moments $\nu_{3j} = E\{|Y_j - E(Y_j)|^3\}$ exist, implying the existence of $\sigma_j^2 = \text{var}(Y_j)$, and that

$$\sum_{j=1}^{n} \nu_{3j} \bigg/ \left(\sum_{j=1}^{n} \sigma_j^2\right)^{3/2}$$

tends to zero as $n \to \infty$, these being called the Liapunov conditions.

Show that the conditions (ii) are satisfied if the $\{Y_j\}$ are uniformly bounded and $\sum\sigma_j^2$ is divergent.

Construct an example, for instance involving exponentially distributed random variables of unequal means, where the variances are finite, but asymptotic normality does not hold.

[Section 2.3; Renyi, 1970, p. 442]

2.4 If the means and variances of a sequence of random variables converge to limits and if all cumulants above a certain order tend to zero, then the sequence is asymptotically normal. That is, information about the remaining cumulants is not required. This result has applications to the theory of random graphs.

[Section 2.3; Janson, 1988]

2.5 A stationary first-order linear autoregressive process of zero mean and lag one correlation ρ, $|\rho| < 1$, is defined by

$$Y_{s+1} = \rho Y_s + Z_{s+1} \qquad (s = 0, \pm 1, \ldots),$$

where $\{Z_s\}$ is a sequence of independent and identically distributed random variables, copies of a random variable Z of zero mean and cumulants $\kappa_r(Z)$ $(r = 2, \ldots)$. Show that in equilibrium, i.e. after a long time,

$$Y_n = \sum_{v=0}^{\infty} \rho^v Z_{n-v},$$

and that

$$K(Y_n; t) = \sum_{v=0}^{\infty} K(Z; \rho^v t),$$

so that in particular $\operatorname{var}(Y_n) = \kappa_2(Z)/(1-\rho^2), \kappa_3(Y_n) = \kappa_3(Z)(1-\rho^3)$. Show that in order that the standardized third cumulant of Y_n is much smaller than that of $\{Z\}$, assumed nonzero, it is necessary that $|\rho|$ is quite large.

Suppose that Y_0 has the equilibrium distribution determined above and let $\bar{Y}_n = (Y_1 + \cdots + Y_n)/n$. By expressing $\bar{Y}_n$ in terms of the $\{Z_s\}$, obtain the standardized third cumulant of $\bar{Y}_n$. Discuss in particular the case of small $|\rho|$ by expanding the standardized cumulant as far as the term in ρ^2.

[Section 2.3]

2.6 Prove carefully from first principles that o, O and o_p, O_p obey the same rules, as evidenced by

$$O(n^{-a})O(n^{-b})=O(n^{-a-b}), \qquad O_p(n^{-a})O(n^{-b})=O_p(n^{-a-b}),$$
$$O_p(n^{-a})O_p(n^{-b})=O_p(n^{-a-b}), \qquad O(n^{-a})o(n^{-b})=o(n^{-a-b}),$$
$$O_p(n^{-a})o(n^{-b})=o_p(n^{-a-b}), \qquad o_p(n^{-a})O(n^{-b})=o_p(n^{-a-b}),$$
$$O_p(n^{-a})o_p(n^{-b})=o_p(n^{-a-b}), \qquad o(n^{-a})o(n^{-b})=o(n^{-a-b}),$$
$$o_p(n^{-a})o(n^{-b})=o_p(n^{-a-b}), \qquad o_p(n^{-a})o_p(n^{-b})=o_p(n^{-a-b}).$$

[Section 2.3]

2.7 Show that if $X_n \xrightarrow{\mathcal{D}} X$, then $\liminf \mathrm{var}(X_n) \geqslant \mathrm{var}(X)$, where if a variance is divergent it is set equal to $+\infty$.

[Section 2.3]

2.8 Prove the result (2.22) that if $U_n \xrightarrow{\mathcal{D}} F(u)$ and $V_n \xrightarrow{p} c > 0$, then $W_n = U_n V_n \xrightarrow{\mathcal{D}} F(w/c)$. To do this evaluate $P(U_n V_n \leqslant w)$ dealing separately with the cases where V_n is or is not in a small neighbourhood of c.

[Section 2.4]

2.9 Let $\{Y_1, Y_2, \ldots\}$ be a sequence of independent and identically distributed random variables, and let $V_n(Y_1, \ldots, Y_n)$ be a function of $Y_1, \ldots, Y_n$. Let

$$\tilde{V}_n = \sum_{j=1}^{n} E(V_n | Y_j) - (n-1)E(V_n),$$

called the Hájek projection. Prove that $E(\tilde{V}_n) = E(V_n)$ and that

$$E(\tilde{V}_n - V_n)^2 = \mathrm{var}(V_n) - \mathrm{var}(\tilde{V}_n).$$

Deduce that if $\mathrm{var}(\tilde{V}_n)/\mathrm{var}(V_n) \to 1$ and $\{\tilde{V}_n - E(\tilde{V}_n)\}/\sqrt{\mathrm{var}(\tilde{V}_n)}$ has a limiting distribution as $n \to \infty$, then $\{V_n - E(V_n)\}/\sqrt{\mathrm{var}(V_n)}$ has the same limiting distribution.

[Section 2.4; Hájek, 1968]

2.10 Let $\{Y_1, Y_2, \ldots\}$ be a sequence of independent and identically distributed random variables, let $\phi_m(x_1, \ldots, x_m)$ be a function of m arguments, symmetrical in the arguments, and write

$$U_n = \sum_{1 \leqslant j_1 < \cdots < j_m \leqslant n} \phi_m(Y_{j_1}, \ldots, Y_{j_m}) \bigg/ \binom{n}{m},$$

so that U_n is an average of values of ϕ_m over all distinct choices of argument from the first n Y's. Then U_n is called a U-statistic of degree m and kernel ϕ. Show via Hájek's projection lemma of the previous exercise that $(U_n - \theta)\sqrt{n}$ is asymptotically normal with zero mean and variance ω^2, where

$$\theta = E\{\phi_m(Y_1, \ldots, Y_m)\}, \qquad \omega^2 = E[E\{\phi(Y_1, \ldots, Y_m | Y_1)\} - \theta]^2,$$

provided that $\omega > 0$; if $\omega = 0$ a different normalization is required and a nonnormal limit may result.

Examine the special cases $\phi_2(x_1, x_2) = (x_1 - x_2)^2, |x_1 - x_2|$.

[Section 2.4; Hoeffding, 1948; van Zwet, 1984]

2.11 Examine the results obtained from Example 2.11 by putting $Z = \log X$.

[Section 2.6]

2.12 Let $\{Y_1, Y_2, \ldots\}$ be independent and identically distributed scalar random variables, copies of a random variable Y with mean μ, variance κ_2 and standardized cumulants ρ_r. Show that if $\mu = 0$, the covariance matrix of $(Y, Y^2)^{\mathrm{T}}$ is

$$\Gamma = \begin{bmatrix} \kappa_2 & \rho_3 \kappa_2^{3/2} \\ \rho_3 \kappa_2^{3/2} & (2 + \rho_4)\kappa_2^2 \end{bmatrix}.$$

Hence show that $(M_{1n}, M_{2n}^{\dagger})$, where

$$M_{1n} = (Y_1 + \cdots + Y_n)/n, \qquad M_{2n}^{\dagger} = \{(Y_1 - \mu)^2 + \cdots + (Y_n - \mu)^2\}/n$$

is asymptotically bivariate normal with mean (μ, κ_2) and covariance matrix Γ/n. Show that the same conclusion holds if $M_{2n}^{\dagger}$ is replaced by $M_{2n} = \Sigma(Y_j - M_{1n})^2/n$.

Show further that if Y is a positive random variable,

$$\sqrt{n}\{(M_{2n}/M_{1n}, M_{1n}) - (\kappa_2/\mu, \mu)\} \simeq \mathrm{MN}_2(0, \Gamma_1),$$

where

$$\Gamma_1 = \begin{bmatrix} \{(2 + \rho_4)\kappa_2^2 \mu^2 + \kappa_2^3 - 2\mu\rho_3\kappa_2^{5/2}\}/\mu^4 & \rho_3\kappa_2^{3/2}/\mu - \kappa_2^2/\mu^2 \\ \rho_3\kappa_2^{3/2}/\mu - \kappa_2^2/\mu^2 & \kappa^2 \end{bmatrix}.$$

Examine the special case where Y has a Poisson distribution. Verify the vanishing of the off-diagonal element in Γ_1 by a conditional argument.

[Sections 2.7, 2.9]

2.13 Develop the discussion corresponding to Example 2.16 taking the reciprocal of a binomial random variable rather than the reciprocal of a Poisson random variable.

[Section 2.8; Grab and Savage, 1954; Tiku, 1964]

2.14 Study the effect of nonnormality on the expansion (2.49) for the expected value of the usual estimate of a standard deviation. For this use the results that if $W_n = \sum(Y_i - \bar{Y}_n)^2/(n-1)$ is the unbiased estimate of variance from independent and identically distributed random variables $Y_1, \ldots, Y_n$, then the first four cumulants of W_n are $\sigma^2, \sigma^4\{\rho_4/n + 2/(n-1)\}$, $\sigma^6[\rho_6/n^2 - 12\rho_4/\{n(n-1)\} + 4\rho_3^2(n-2)/\{n(n-1)^2\} + 8/(n-1)^2]$, $\sigma^8\{\rho_8/n^3 + 24\rho_6/\{n^2(n-1)\} + 32\rho_5\rho_3(n-2)/\{n(n-1)\}^2 + 8\rho_4^2(4n^2 - 9n + 6)/\{n^2(n-1)^3\} + 144\rho_4/\{n(n-1)^2\} + 96\rho_3^2/\{n(n-1)^3\} + 48/(n-1)^3]$, where σ^2 and the ρ's refer to the distribution of Y.

[Section 2.8]

2.15 Let f_n and f be probability density functions of p-dimensional random variables X_n and X; show that if $f_n(x) \to f(x)$ as $n \to \infty$ for almost all (with respect to Lebesgue measure) $x \in R^m$ then $X_n \overset{\mathscr{D}}{\to} X$.

[Section 2.9]

2.16 Show that the concept of weak convergence remains the same if in Definition 2.1 one substitutes $\bar{C}_0(R^m)$ for $\bar{C}(R^m)$ where $\bar{C}_0(R^m)$ denotes the set of bounded and uniformly continuous functions on R^m.

[Section 2.9]

2.17 The result of the previous exercise may be used to prove Theorem 2.2.

[Section 2.9]

2.18 Generalize Theorem 2.6 to the case $c_n(U_n - u_0) \overset{\mathscr{D}}{\to} W$ for some random variable W and some sequence of constants $c_n \to \infty$.

[Section 2.9]

Bibliographic notes

Most of the ideas in section 2.3 are described in detail in any medium-level book on the theory of probability. See, for example, Gnedenko (1962), Moran (1968) and Renyi (1970). Mann and Wald (1943) introduces the o_p, O_p notation; see Bishop, Fienberg and

Holland (1975) for a detailed account. The use of local linearization to obtain approximate variances of nonlinear functions goes back to Gauss. The associated argument for finding transformations with constant variance probably dates from Fisher's (1921) transformation of the correlation coefficient; see Anscombe (1948) and for further mathematical discussion, Hougaard (1982). For a thorough discussion of extreme value distributions, see Galambos (1978).

CHAPTER 3

Asymptotic expansions

3.1 Introduction

It is convenient to begin the more formal discussion of asymptotic expansions by a brief account of such expansions for a nonrandom sequence or function.

Let $\{b_{kn}\}$ for $k = 0, 1, \dots$ be a base set of sequences such that as $n \to \infty$, $b_{kn} = o(b_{k-1,n})$ with $b_{0n} = 1$. Typical examples are

$$b_{0n} = 1, \quad b_{1n} = 1/\sqrt{n}, \quad b_{2n} = 1/n, \dots;$$
$$b_{0n} = 1, \quad b_{1n} = 1/n, \quad b_{2n} = 1/n^2, \dots,$$

although occasionally more complicated sequences are needed, as for example when discussing stable laws.

For a given sequence $\{g_n\}$, suppose that there is a sequence $\{g_n^{(0)}\}$, usually of rather simple form, such that as $n \to \infty$ for each fixed $h = 0, 1, \dots$ and for suitable constants $\gamma_1, \gamma_2, \dots,$

$$g_n = g_n^{(0)}\{1 + \gamma_1 b_{1n} + \cdots + \gamma_h b_{hn} + O(b_{h+1,n})\}, \tag{3.1}$$

so that in particular

$$g_n = g_n^{(0)}\{1 + O(b_{1n})\}. \tag{3.2}$$

Then we call (3.1) an *asymptotic expansion* of $\{g_n\}$ as $n \to \infty$ with *leading term* $g_n^{(0)}$.

In the applications with which we shall be concerned, we take normally the leading term with at most one or two further terms, i.e. we are concerned with the general definition (3.1) only for $h = 0, 1, 2$. In particular we are not concerned with the convergence of the infinite series (3.1) as $h \to \infty$ for fixed n. Note also that by considering expansions for $g_n/g_n^{(0)}$ we could without essential loss of generality for much of the discussion take the leading term to be one.

A similar set of definitions applies for the asymptotic expansion of a function $g(z)$ as, say, $z \to \infty$, the leading term being a function

$g^{(0)}(z)$ and the base set of sequences being replaced by a base set of functions $b_0(z), b_1(z), \ldots$ such that $b_0(z) = 1$ and as $z \to \infty$

$$b_{h+1}(z) = o(b_h(z));$$

we require that

$$g(z) = g^{(0)}(z)\{1 + \gamma_1 b_1(z) + \cdots + \gamma_h b_h(z) + O(b_{h+1}(z))\}. \qquad (3.3)$$

Similar definitions apply if we are interested in some other limiting value of z.

In probabilistic and statistical applications the quantity becoming large is usually a sample size or an amount of information and is conveniently denoted by n. We shall make this notational change later but for the moment retain z as suggestive of a real or complex variable.

Operations on asymptotic expansions, e.g. addition and multiplication, can be carried out in an obvious way, as can term-by-term integration, subject to simple conditions on the convergence of the integrals involved. Differentiation term by term requires that the differentiated series should be an asymptotic expansion. We shall not concern ourselves with rigorous general formulations.

It quite often happens that the functions to be expanded have further variables or parameters, θ say, which are to be regarded as fixed as $z \to \infty$. It is then important to specify the region of θ for which the asymptotic expansion holds. Sometimes, however, as θ approaches particular values the asymptotic property of the expansion with respect to z fails. It may then be desirable to consider 'matched' asymptotic series in which, in effect, new 'coordinates' are introduced in the region of failure of the main expansion. We shall not consider this in detail, but we discuss an important special case in section 3.9. Moreover, Example 2.9 is a very simple illustration.

3.2 Integration by parts

The following simple examples illustrate the use of integration by parts and are intended both to be of intrinsic interest and to exemplify general principles.

Example 3.1 Normal integral

To examine the behaviour of the standardized normal integral for large arguments, let $\phi(y) = (2\pi)^{-1/2} e^{-1/2y^2}$ and consider for large z

$$\Psi(z) = \int_z^\infty \phi(y) dy. \tag{3.4}$$

Then

$$\begin{aligned}\Psi(z) &= -\frac{1}{\sqrt{(2\pi)}} \int_z^\infty \frac{d(e^{-1/2y^2})}{y} \\ &= \frac{\phi(z)}{z} - \int_z^\infty \frac{\phi(y)dy}{y^2}.\end{aligned} \tag{3.5}$$

Repetition of this process yields

$$\Psi(z) = \frac{\phi(z)}{z}\left\{1 - \frac{1}{z^2} + \frac{3}{z^4} - \frac{3 \cdot 5}{z^6} + \cdots\right\}. \tag{3.6}$$

The remainder if we stop at the term in z^{-2h} is easily shown to be $O(z^{-2h-2})$; for instance, if $h = 0$, so that just the leading term is taken, the remainder is the second term of (3.5).

The following properties follow quite directly:

1. Equation (3.6) defines an asymptotic expansion as $z \to \infty$ with leading term $\phi(z)/z$ and base functions $\{1, 1/z^2, 1/z^4, \ldots\}$;
2. The expansion is *strictly alternating*: not only do the terms alternate in sign but the successive partial sums provide alternatively upper and lower bounds to the target function $\Psi(z)$;
3. For fixed z, the infinite series (3.6) is divergent;
4. For fixed z, the error committed in terminating (3.6) is less than the first omitted term;
5. Table 3.1 illustrates numerically the nature of the approximations. For convenience $z\Psi(z)/\phi(z)$ has been taken as the function under study, so that the leading term is 1;
6. For any given z it is dangerous to take too many terms; note especially what happens at $z = 1$, where, because $S_1, S_3, \ldots$ are upper bounds and $S_3 > S_1$, it follows immediately that three terms are worse than one;
7. For a given number r of terms, S_r improves as z increases: this is to be expected in a general way from the defining property of an asymptotic expansion, although the monotonic improvement cannot be expected universally;
8. For any given z, not too small, the average, S^*, of the best upper bound and the best lower bound is a rather good approximation;

Table 3.1 *Comparison of $z\Psi(z)\phi(z)$ with S_r, sum of r terms of asymptotic expansion (3.6).*

z	$z\Psi(z)\phi(z)$	S_1	S_2	S_3	S_4	S^*
1	0·6551	1	0	3	−12	0·5
2	0·8427	1	0·75	0·9375	0·7031	0·8438
3	0·9131	1	0·8889	0·9259	0·9053	0·9133
4	0·9466	1	0·9375	0·9492	0·9456	0·9466
5	0·9640	1	0·96	0·9648	0·9638	·
10	0·9903	1	0·99	0·9903	0·9903	·

S^*, average of best upper and best lower bound

9. The expansion, while important for analytical purposes, is not much use for direct statistical work in which interest is focused largely on the range $|z| \leqslant 3$;
10. Different functions can have the same asymptotic expansion. For instance $\Psi(z) + \exp(-z^4)$ has the asymptotic expansion (3.6), because, as $z \to \infty$, the added term is very small compared with all the terms in (3.6).

Example 3.2 Incomplete gamma function

A very similar argument applies to the incomplete gamma function, which it is convenient to write in the form, for $m > 0$,

$$\gamma_m(z) = \int_z^\infty y^{m-1} e^{-y} dy / \Gamma(m)$$

$$= \frac{z^{m-1} e^{-z}}{\Gamma(m)} + \frac{m-1}{\Gamma(m)} \int_z^\infty y^{m-2} e^{-y} dy, \tag{3.7}$$

after integration by parts. On repeated integration by parts, we have that

$$\gamma_m(z) = \frac{z^{m-1} e^{-z}}{\Gamma(m)} \left\{ 1 + \frac{m-1}{z} - \frac{(m-1)^{[2]}}{z^2} + \cdots \right\}, \tag{3.8}$$

where $(m-1)^{[r]} = (m-1)(m-2)\cdots(m-r)$. The error in stopping at the term in $(m-1)^{[h]} z^{-h}$ is

$$\frac{(m-1)^{[h+1]}}{\Gamma(m)} \int_z^\infty y^{m-h-2} e^{-y} dy. \tag{3.9}$$

Now if $h > m-2$, (3.9) is less in absolute value than

$$|(m-1)^{[h+1]}|z^{m-h-2}e^{-z}/\Gamma(m),$$

the next term in the series, and the error has the same sign as the next term. For smaller values of h a series of lower bounds is obtained. For integer m, the series is exact after a suitable finite number of terms.

Note that with $m=\frac{1}{2}$, $\gamma_{1/2}(z)=\Psi(\sqrt{(2z)})$ in the notation of Example 3.1 and in that sense the present example is a generalization of the earlier one. The probabilistic interpretation comes via (3.7) considered as the survivor function of one half chi-squared with one degree of freedom.

Example 3.3 Exponential integral distribution

Let Y be the interval from the time origin to the first point in a Poisson process of unit rate, so that Y has density e^{-y} $(y \geq 0)$. Let a sampling point be placed at random uniformly over the interval and let Z be its distance from the origin. Then Z is easily shown to have density

$$\int_z^\infty \frac{e^{-y}}{y}\,dy = E_1(z), \tag{3.10}$$

say, and cumulative distribution function

$$1-e^{-z}+zE_1(z).$$

The function $E_1(z)$ is called the exponential integral and arises in other contexts too (Abramowitz and Stegun, 1965).

Now

$$E_1(z)=\lim_{m\to 0}\{\Gamma(m)\gamma_m(z)\}$$

and the argument of Example 3.2 produces the strictly alternating asymptotic expansion

$$E_1(z)=\frac{e^{-z}}{z}\left(1-\frac{1}{z}+\frac{2!}{z^2}-\frac{3!}{z^3}+\cdots\right). \tag{3.11}$$

3.3 Laplace expansion

Some unity is brought to the examples of Section 3.2 by considering the expansion for large z of the Laplace transform

$$\mathscr{G}(z) = \int_0^\infty e^{-zy} g(y) dy. \tag{3.12}$$

For well-behaved functions $g(y)$ the form of $\mathscr{G}(z)$ for large z is determined by the values of $g(y)$ near $y = 0$, an instance of an Abelian theorem. Repeated integration by parts yields as $z \to \infty$ the expansion

$$\mathscr{G}(z) = g(0)/z + g'(0)/z^2 + \cdots. \tag{3.13}$$

The defining property of an asymptotic expansion is proved by examining the remainder after $k + 1$ integration by parts; in particular cases the study of the remainder may allow a bound on error to be calculated. The expansion is derivable also by expanding $g(y)$ in the Taylor series

$$g(y) = \sum g^{(r)}(0) y^r / r!$$

and integrating term by term. This can be justified very directly from a special case of Watson's lemma; see Further results and exercises, 3.5.

The condition for (3.13) to be a strictly alternating series is that the derivatives of $g(y)$ at $y = 0$ alternate in sign, in which case $g(y)$ is called *absolutely monotone.*

Note that the range of integration in (3.12) can be changed to $(0, a)$ for any fixed $a > 0$ without affecting (3.13).

The qualitative idea involved here is not restricted to (3.12) but applies whenever an integral, or indeed a sum, over a substantial range is largely determined by behaviour near one (or a few) particular points.

We consider the behaviour as $z \to \infty$ of

$$w(z) = \int_a^b e^{-zr(y)} g(y) dy. \tag{3.14}$$

For large $z, w(z)$ is determined by behaviour near the minimum of $r(y)$, at $y = \tilde{y}$, say, although, of course, we need $g(\tilde{y}) \neq 0$ for a useful answer to emerge.

Note that (3.14) is not the most general form that can be handled by the argument that follows: if $g(y)$ depends on z and can, say, be expanded in powers of $1/\sqrt{z}$, it will usually be possible to proceed term by term and, as we shall see later, mild dependence of $r(y)$ on z can also be accommodated.

There are two broad possibilities to be considered for (3.14), each with relatively minor variants:

1. the minimum of $r(y)$ is at a or b and $r'(y)$ and $g(y)$ are nonzero there;
2. the minimum of $r(y)$ is at an interior point $y = \tilde{y}$, say, where $r'(\tilde{y}) = 0$, $r''(\tilde{y}) > 0$, $g(\tilde{y}) \neq 0$.

Case (1) is a fairly direct extension of the special case of the Laplace transform $a = 0$, $r(y) = y$. While a formulation embracing all or most cases is possible, we shall concentrate on case (2). In outline we write

$$w(z) = \int_a^b \exp\{-zr(\tilde{y}) - z(y-\tilde{y})r'(\tilde{y}) - \tfrac{1}{2}z(y-\tilde{y})^2 r''(\tilde{y}) - \cdots\}g(y)dy \tag{3.15}$$

and because $r'(\tilde{y}) = 0$, the quadratic term in the exponent is the key one and it is natural to make the transformation

$$(y-\tilde{y})\sqrt{z}\sqrt{\tilde{r}''} = v,$$

from which follows an expansion in powers of $1/\sqrt{z}$ with leading term

$$\frac{e^{-z\tilde{r}}\tilde{g}\sqrt{(2\pi)}}{\sqrt{z}\sqrt{\tilde{r}''}}\{1 + O(z^{-1})\}, \tag{3.16}$$

where $\tilde{r} = r(\tilde{y}), \tilde{g} = g(\tilde{y}), \tilde{r}'' = r''(\tilde{y})$. Detailed calculation shows that the odd-order terms in $1/\sqrt{z}$ vanish.

Direct if tedious expansion of (3.15) produces the further expansion

$$w(z) = \frac{e^{-z\tilde{r}}\sqrt{(2\pi)}}{\sqrt{z}\sqrt{\tilde{r}''}}\left\{\tilde{g} + \frac{1}{z}\left(\frac{\tilde{g}''}{2\tilde{r}''} - \frac{\tilde{r}^{(3)}\tilde{g}'}{2(\tilde{r}'')^2} - \frac{\tilde{r}^{(4)}\tilde{g}}{8(\tilde{r}'')^2} + \frac{5(\tilde{r}^{(3)})^2\tilde{g}}{12(\tilde{r}'')^3}\right) + O(z^{-2})\right\}. \tag{3.17}$$

It will help to establish the connexion with our probabilistic methods to write the integrand of (3.14) as

$$g(y)e^{-z\tilde{r}}\phi\{y-\tilde{y}; (z\tilde{r}'')^{-1}\}/\phi\{0; (z\tilde{r}'')^{-1}\},$$

where $\phi(x;\kappa)$ is the density of a normal distribution of zero mean and variance κ, so that in particular $\phi(0;\kappa) = (2\pi\kappa)^{-1/2}$.

We can then interpret (3.16) approximately as

$$w(z) = \frac{e^{-z\tilde{r}}}{\phi\{0; (z\tilde{r}'')^{-1}\}}E\{g(Y_z)\}, \tag{3.18}$$

where Y_z is a random variable normally distributed with mean $\tilde{r}$ and variance $(z\tilde{r}'')^{-1}$, and where the region of integration in (3.14) has effectively been transformed to $(-\infty, \infty)$.

Note that if we make a preliminary transformation of y in (3.14) to a new variable t so that $r(y) = r(\tilde{y}) + \frac{1}{2}(t - \tilde{t})^2$, then (3.18) is exact when the interval of integration (a, b) is the whole line, $g(y)$ being replaced by $g(y)dy/dt$ expressed as a function of t. It is clear from this that the argument applies equally to integrals of the form

$$w(z) = \int_a^b e^{-zr_z(y)} g(y)dy \tag{3.19}$$

provided that, for suitable choice of the defining parameter z, we can write near the minimum $y = \tilde{y}_z$:

$$\frac{e^{-zr_z(y)}}{e^{-zr_z(\tilde{y}_z)}} = \frac{\phi\{y - \tilde{y}_z; (z\tilde{\omega})^{-1}\}}{\phi\{0; (z\tilde{\omega})^{-1}\}} \{1 + O(z^{-1/2})\}. \tag{3.20}$$

In particular, if the position of the minimum of $r_z(y)$ depends on z but the second derivative is a fixed multiple of z, the argument is essentially unchanged; in other cases an asymptotic expansion of the second derivative at the minimum may be needed.

The method producing (3.16) and similar results is named after Laplace.

In the first form outlined here it is required that the base point $\tilde{y}$ is independent of z. Some preliminary manoeuvring may be needed to achieve this if the function under study is not initially exactly in the form (3.14). It is, as noted above, possible to allow mild dependence of the base point on z, such as would follow for instance if we rewrote (3.14) in the form

$$w(z) = \int \exp\{-zr(y) + \log q(y)\} dy \tag{3.21}$$

and expanded about the maximum of the whole exponent. That is, we write

$$z\rho(y, z) = zr(y) - \log q(y)$$

and denote the position of the minimum by y^*, which will depend (slightly) on z. We write

$$\rho^{*''} = \partial^2 \rho(y, z)/\partial y^2$$

evaluated at $y = y^*$, with a similar notation for higher derivatives.

Then application of (3.20) gives

$$w(z)=\frac{e^{-z\rho^*}\sqrt{(2\pi)}}{\sqrt{z}\sqrt{\rho^{*''}}}\left[1+\frac{1}{z}\left\{-\frac{\rho^{*(4)}}{8(\rho^{*''})^2}+\frac{5(\rho^{*(3)})^2}{12(\rho^{*''})^3}\right\}+O(z^{-2})\right]. \tag{3.22}$$

It is not possible to say in general when the leading term of (3.22) is preferable to the simpler leading term of (3.17), although if $q(y)$ varies strongly with y in the neighbourhood of $y=\tilde{y}$ it seems likely that the second form will be the more accurate. (This second form is important for the discussion of posterior expectations in Example 3.9.) Thus for

$$\int_{-\infty}^{\infty} e^{-1/2ny^2}\{1+g_1(y)\}\,dy,$$

where $g_1(y)=-g_1(-y)$ and $|g_1(y)|<1$ it is clear that incorporation of the second factor into the exponent is disadvantageous, whereas, at the other extreme, for

$$\int_{-\infty}^{\infty} e^{-1/2ny^2}e^{1/2y^2}\,dy$$

such incorporation gives the exact answer.

Example 3.4 Normal integral (continued)

Example 3.1 is not immediately of the form considered in the present section. It can, however, be made so by a simple change of variables $y=z+x$, as follows:

$$\begin{aligned}\Psi(z)&=\int_z^{\infty}\phi(y)\,dy\\ &=\int_0^{\infty} e^{-zx}e^{-1/2x^2}\,dx\,\frac{e^{-1/2z^2}}{\sqrt{(2\pi)}}\end{aligned} \tag{3.23}$$

and the previous expansion follows via (3.13), being a consequence of the domination of the integral in (3.23) by behaviour near $x=0$.

Example 3.5 Stirling's theorem

We now examine the behaviour for large z of

$$\Gamma(z+1)=\int_0^{\infty} x^z e^{-x}\,dx.$$

For fixed z the maximum of the integrand is at $x = z$ and it is natural in the light of the above discussion to transform to make the maximum at a fixed point, and the obvious way to do this is to write $y = x/z$. Then

$$\Gamma(z+1) = z^{z+1} \int_0^\infty \exp(z \log y - zy) dy \tag{3.24}$$

and this is exactly of the form (3.14) with $\tilde{y} = 1$, $r(y) = -\log y + y$, $g(y) = 1$. We can either substitute in (3.20) or develop the corresponding expansions from first principles to obtain

$$\Gamma(z+1) = \sqrt{(2\pi)} z^{z+1/2} e^{-z} \left\{ 1 + \frac{1}{12z} + O(z^{-2}) \right\}. \tag{3.25}$$

We can now illustrate the useful idea of amalgamation into the leading term. The asymptotic expansion of a function to a given order of accuracy can be rewritten in many forms agreeing to the order in question. Having obtained such an expansion in a particular version, we can sometimes usefully rewrite the expansion differently, so as in effect to incorporate the higher-order terms into the leading term. The reason for this may be to obtain a more elegant version of the expansion convenient for further analytical work or to improve the numerical adequacy, in particular by extending the range of applicability.

Example 3.6 Stirling's theorem (continued)

We start from (3.25) and wish to amalgamate $1/(12z)$ into the leading term. One reason for this is to improve accuracy for small z, in particular at or near $z = 0$, where (3.25) fails completely. To do this note that we can write for any fixed a,

$$\Gamma(z+1) = (2\pi)^{1/2} z^{z+1/2-a} e^{-z} \{z + 1/(12a)\}^a \{1 + O(z^{-2})\},$$

because expansion of the first factor reproduces (3.25). One way of choosing a is to produce agreement with the leading term at $z = 0$, leading to $(2\pi)^{1/2} = (12a)^a$, the solution of which is close to $a = \frac{1}{2}$. This leads to the simple version

$$\Gamma(z+1) = (2\pi)^{1/2} z^z e^{-z} (z + 1/6)^{1/2} \{1 + O(z^{-2})\}. \tag{3.26}$$

At $z = 0$ the leading term of this is 1·023, which for many purposes is entirely adequate agreement.

In some ways a more systematic approach is to attempt to 'match' the asymptotic expansion (3.25) for large z with the expansion $\Gamma(z+1)=(z+1)^{-1}+\cdots$ holding near the singularity at $z=-1$. For this we write

$$\Gamma(z+1)=b(z+1)^{-1}(z+c)^{z+d}e^{-z}\{1+O(z^{-2})\}.$$

Expansion for large z shows that the leading term is

$$bz^{z+d-1}e^{-z}z^{c}\{1+(dc-\tfrac{1}{2}c^2-1)/z+O(z^{-2})\}$$

and there are now various ways of choosing the disposable constants b, c, d. If we choose the constants to match the original expansion we require that

$$c=3/2+3/6 \doteq 1{\cdot}7887, \quad b=(2\pi)^{1/2}e^{-1} \doteq 0{\cdot}4191, \quad d=3/2,$$

giving

$$\Gamma(z+1)=0{\cdot}4191(z+1)^{-1}(z+1{\cdot}7887)^{z+1/2}e^{-z}. \qquad (3.27)$$

As $z \to -1+$, this behaves like $1{\cdot}012/(z+1)$, almost reproducing the required behaviour. Obviously there are other ways in which b, c, d can be chosen.

Table 3.2 compares the accuracy of Stirling's formula and the various modifications discussed here. Of course to confront a formula developed for large z with the behaviour near $z=0$ is a severe test

Table 3.2 *Stirling's formula and simple modifications.*

z	$\Gamma(z+1)$	*Leading term* (3.25)	*Two terms* (3.25)	(3.26)	(3.27)
−0·9	9·5135	·	·	·	9·6036
−0·8	4·5908	·	·	·	4·6267
−0·6	2·2182	·	·	·	2·2305
−0·4	1·4892	·	·	·	1·4954
−0·2	1·1642	·	·	·	1·1680
0	1	2·5066	·	1·0233	1·0026
0·5	0·8862	0·7602	0·8869	0·8778	0·8877
1	1	0·9221	0·9990	0·9960	1·0011
1·5	1·3293	1·2584	1·3283	1·3265	1·3305
2	2	1·9190	1·9990	1·9974	2·0014
3	6	5·8362	5·9983	5·9961	6·0028

and it is remarkable how well the formula performs even without the special modifications.

Example 3.7 Digamma function: the log gamma distribution

The kth derivative of the gamma function is given by

$$\Gamma^{(k)}(z+1) = \int_0^\infty y^z (\log y)^k e^{-y}\,dy. \tag{3.28}$$

Asymptotic expansions can be obtained for a fixed integer k and large z by arguing from first principles or by differentiating Stirling's formula; conditions on the differentiated series have to be verified for this to be valid. In fact, by exploiting special properties of the gamma function we can argue slightly more directly and also give the calculations a probabilistic interpretation, although of course, this last is not essential.

For such a probabilistic interpretation, it is convenient to make a small change in notation, writing $m = z + 1$ and denoting by W_m a random variable having density $w^{m-1}e^{-w}/\Gamma(m)$.

Then

$$\Gamma^{(k)}(m)/\Gamma(m) = E\{(\log W_m)^k\}. \tag{3.29}$$

We standardize by writing $W_m^* = (W_m - m)/\sqrt{m}$, so that

$$\begin{aligned}\Gamma^{(k)}(m)/\Gamma(m) &= E\{\log(m + W_m^*\sqrt{m})\}^k \\ &= E\{\log m + \log(1 + W_m^*/\sqrt{m})\}^k,\end{aligned}$$

leading to an expansion in powers of $1/\sqrt{m}$ with coefficients determined by the moments of W_m^*.

For simplicity we concentrate on the case $k = 1$ when

$$\Gamma'(m)/\Gamma(m) = E\left(\log m + \frac{W_m^*}{\sqrt{m}} - \frac{W_m^{*2}}{2m} + \frac{W_m^{*3}}{3m\sqrt{m}} - \frac{W_m^{*4}}{4m^2} + \cdots\right). \tag{3.30}$$

Now it is easily shown, for example from the cumulant generating function of W_m, that

$$E(W_m^*) = 0, \quad E(W_m^{*2}) = 1, \quad E(W_m^{*3}) = 2/\sqrt{m}, \quad E(W_m^{*4}) = 3 + 6/m,$$

so that

$$\Gamma'(m)/\Gamma(m) = \log m + \frac{1}{6m} + O(m^{-2}). \tag{3.31}$$

Again, by amalgamation into the leading term this can be written as

$$\Gamma'(m)/\Gamma(m) = \log(m + \tfrac{1}{6}) + O(m^{-2}).$$

It is instructive to compare the expansion involved in the above argument with that involved in a direct application of Laplace's method, as in (3.25). Essentially the special form of $r(y)$ has obviated the need to expand that function.

Example 3.8 The von Mises distribution

The von Mises distribution for angles on a circle has density defined for $-\pi \leqslant y < \pi$ by

$$\frac{e^{\kappa \cos(y-\alpha)}}{2\pi I_0(\kappa)}, \tag{3.32}$$

where κ is a concentration parameter, $\kappa > 0$, and α is a location parameter. For our purpose we may take $\alpha = 0$. The normalizing constant $I_0(\kappa)$ is a Bessel function of zero order and imaginary argument and is defined by

$$I_0(\kappa) = \frac{1}{2\pi}\int_{-\pi}^{\pi} e^{\kappa \cos y}\, dy. \tag{3.33}$$

For small κ, (3.33) can be studied by expanding $e^{\kappa \cos y}$ in powers of κ and integrating term by term. We examine behaviour for large κ. The integrand in (3.33) has its maximum at $y = 0$ and so is directly in a form for the application of Laplace's method with $\tilde{y} = 0$. We write $x = y\sqrt{\kappa}$, leading to

$$\begin{aligned}
I_0(\kappa) &= \frac{1}{2\pi}\int_{-\pi\sqrt{\kappa}}^{\pi\sqrt{\kappa}} \exp\{\kappa \cos(x/\sqrt{\kappa})\}\frac{dx}{\sqrt{\kappa}} \\
&= \frac{1}{2\pi\sqrt{\kappa}}\int_{-\pi\sqrt{\kappa}}^{\pi\sqrt{\kappa}} \exp\left\{\kappa - \frac{x^2}{2} + \frac{x^4}{24\kappa} + \cdots\right\} dx \\
&= \frac{e^{\kappa}}{2\pi\sqrt{\kappa}}\int_{-\infty}^{\infty} e^{-1/2x^2}\left\{1 + \frac{x^4}{24\kappa} + O(\kappa^{-3/2})\right\} dx \\
&= \frac{e^{\kappa}}{\sqrt{(2\pi\kappa)}}\left\{1 + \frac{1}{8\kappa} + O(\kappa^{-3/2})\right\}.
\end{aligned} \tag{3.34}$$

Note that a byproduct of this argument is that as $\kappa \to \infty$ the distribution is asymptotically normal.

It is now convenient to make the change of notation threatened in section 3.1 and to replace z by n.

Example 3.9 Posterior expectations

In a calculation via Bayes's theorem the posterior density of a quantity y of interest, usually an unknown parameter, is often proportional to

$$\exp\{nl(y)+\pi(y)\}, \tag{3.35}$$

where $nl(y)$ is the log likelihood corresponding to some data, for example to n independent and identically distributed random variables, and $\pi(y)$ arises from the prior distribution. For the present discussion we suppose that y is one-dimensional, although the multidimensional application is very important.

Note that the normalizing constant of (3.35) is an integral to which Laplace's method can be applied. Suppose that we require the posterior expectation of some function $g(y)$ of y, this function being independent or almost independent of n. One possibility is $g(y)=y$.

Thus we need the ratio of two integrals, namely

$$\frac{\int \exp\{nl(y)+\pi(y)+h(y)\}\,dy}{\int \exp\{nl(y)+\pi(y)\}\,dy}, \tag{3.36}$$

where the integral in the denominator is the normalizing constant for (3.35) and $h(y)=\log g(y)$. Let the maxima of the two arguments of the exponentials in (3.36) be at $\tilde{y}_{g\pi}$ and $\tilde{y}_{\pi}$ respectively. Then use of the leading term from (3.12) applied separately to numerator and denominator gives

$$\frac{\{-nl''(\tilde{y}_{g\pi})-\pi''(\tilde{y}_{g\pi})-h(\tilde{y}_{g\pi})\}^{1/2}}{\{-nl''(\tilde{y}_{\pi})-\pi''(\tilde{y}_{\pi})\}^{1/2}}$$
$$\times\exp\{nl(\tilde{y}_{g\pi})-nl(\tilde{y}_{\pi})+\pi(\tilde{y}_{g\pi})-\pi(\tilde{y}_{\pi})+h(\tilde{y}_{g\pi})\}. \tag{3.37}$$

This at first sight has error $O(n^{-1})$. In fact, however, from (3.22) the terms of order n^{-1} in numerator and denominator are dominated by the contributions from $nl(y)$ and are the same to order n^{-1}. It follows that (3.37) has error $O(n^{-2})$.

Example 3.10 The sample range

Let $Y_1, \ldots, Y_n$ be independent and identically distributed random variables, copies of a random variable Y with density $f(y)$ and cumulative distribution function $F(y)$. Then the probability density function of the range, $\max(Y_1, \ldots, Y_n) - \min(Y_1, \ldots, Y_n)$, is

$$p_n(x) = n(n-1)\int_{-\infty}^{\infty} f(y-\tfrac{1}{2}x)f(y+\tfrac{1}{2}x)\{F(y+\tfrac{1}{2}x)-F(y-\tfrac{1}{2}x)\}^{n-2}dy \tag{3.38}$$

and Laplace's method can be applied to this for large n and fixed x, with

$$q(y) = f(y-\tfrac{1}{2}x)\,f(y+\tfrac{1}{2}x), \qquad r(y) = \log\{F(y+\tfrac{1}{2}x) - F(y-\tfrac{1}{2}x)\}.$$

There is some simplification if $f(y)$ is a symmetrical density with mode, say, zero. The maximum of $r(y)$ is then at zero and the leading term of the expansion is, after some calculation,

$$\frac{n^{3/2}(2\pi)^{1/2}\{F(\tfrac{1}{2}x)\}^2\{F(\tfrac{1}{2}x) - F(-\tfrac{1}{2}x)\}^{n-3/2}}{\{f'(-\tfrac{1}{2}x)\}^{1/2}}. \tag{3.39}$$

Numerical work shows that the correction term is positive so that for fixed n the integral of (3.39) is less than one. This suggests the semi-empirical adjustment of normalizing (3.39), i.e. of multiplying (3.39) by c_n, depending only on n, determined so that the integral of the approximation is one. We shall use this device again later; note that while in a reasonable sense it produces an improvement on the average, it does not necessarily improve the approximation in particular regions which may be of concern in a specific application.

3.4 Summation of series

We now deal much more briefly with the approximation of sums of series. The most widely used technique applicable when the sum is a large number of terms where the individual terms change relatively smoothly, none being dominant, is to approximate the sum by an integral. For this to work it is in particular necessary that the terms do not rapidly alternate in sign.

The relation between a series and the corresponding integral is expressed in some generality in the Euler–Maclaurin theorem (Olver,

1974, Ch. 8). A simplified version is that

$$S_n = \tfrac{1}{2}f(a) + f(a+1) + \cdots + f(n-1) + \tfrac{1}{2}f(n) \tag{3.40}$$

and

$$I(n) = \int_a^n f(x)dx \tag{3.41}$$

differ by

$$\{f'(n) - f'(a)\}/12 - \{f^{(3)}(n) - f^{(3)}(a)\}/720 + R_4, \tag{3.42}$$

where the remainder R_4 is a certain integral involving $f^{(4)}(x)$. A more general form stops at $f^{(2h-1)}(x)$ and has a remainder R_{2h} depending on $f^{(2h)}(x)$.

While this is not in the form of an asymptotic expansion as $n \to \infty$ it can often be converted into one by suitable manoeuvres.

The proof of (3.42) is by repeated integration by parts, starting from the remark that

$$\int_k^{k+1} f(x)dx = [(x-k-\tfrac{1}{2})f(x)]_k^{k+1} - \int_k^{k+1} (x-k-\tfrac{1}{2})f'(x)dx,$$

so that

$$\tfrac{1}{2}f(k) + \tfrac{1}{2}f(k+1) = \int_k^{k+1} f(x)dx + \int_k^{k+1} \beta_1(x)f'(x)dx,$$

where

$$\beta_1(x) = x - [x] - \tfrac{1}{2}.$$

For full details, see Olver (1974, p. 279).

We illustrate the analytical use of the formula via examples.

Example 3.11 Exponential order statistics

The expected value e_{nr} of the rth order statistic in a random sample of size n from the exponential distribution of unit mean is for $r = 1, \ldots, n$ given by

$$e_{nr} = n^{-1} + \cdots + (n-r+1)^{-1}. \tag{3.43}$$

We take $f(x) = 1/x$ and integrate from $n-r+1$ to n. Some care is needed in applying the Euler–Maclaurin theorem when r is near n because the derivatives at the lower terminal will not be small.

Thus with $r = n$, we have that

$$e_{nn} - \tfrac{1}{2} - (2n)^{-1} = \int_1^n dx/x - \{1/(12n^2) + \cdots\} + A_1,$$

where A_1 is a collection of terms based on the derivatives of $f(x)$ at $x = 1$. Now it is known that as $n \to \infty$

$$e_{nn} - \log n \to \gamma,$$

where $\gamma \doteq 0{\cdot}5772$ is Euler's constant. It follows that $A_1 = \gamma - \tfrac{1}{2}$ and that therefore for large n

$$e_{nn} = \log n + (2n)^{-1} - (12n^2)^{-1} + \gamma. \tag{3.44}$$

For some purposes it is convenient to amalgamate the first few terms in (3.44), replacing the right-hand side by

$$\log\{n + \tfrac{1}{2} + (24n)^{-1}\} + \gamma. \tag{3.45}$$

Note that at $n = 1$, (3.45) is equal to 1·01, the exact value being 1.

For $r < n$, we proceed either by removing a few terms from (3.45) or by applying the Euler–Maclaurin theorem directly to the integral over, say, $(n - r, n)$ to give as an approximation to e_{nr}

$$\log\{n/(n-r)\} + (2n)^{-1} - \{2(n-r)\}^{-1} + \{r(2n-r)\}/\{12n^2(n-r)^2\}.$$

Again this can be amalgamated into

$$\log[\{n + \tfrac{1}{2} + (24n)^{-1}\}/\{n - r + \tfrac{1}{2} + (24n - 24r)^{-1}\}]. \tag{3.46}$$

Of course for many purposes (3.45) and (3.46) can be simplified, in particular the latter to $-\log(1 - r/n)$, corresponding to the log linear form of the survivor function of the exponential distribution.

Example 3.12 Continuity correction

In spirit there is a close connexion between the Euler–Maclaurin theorem and the use of a continuity correction in relating a tail area in a discrete distribution with a corresponding tail area in a continuous distribution.

To explore this relation further, suppose that in (3.40) and (3.41), $f(x)$ vanishes at any lower terminal a and that the same function evaluated at the integer points and for all real x defines a discrete and a continuous distribution. Of course in most applications in which a discrete distribution is approximated by a continuous distribution some further approximation is involved. If we evaluate $F_d(n)$, the discrete distribution function at n, by the Euler–Maclaurin theorem, we have that

$$\begin{aligned} F_d(n) &= \sum_{r \leqslant n} f(r) \\ &\doteq \int_{-\infty}^{n} f(x)dx + \tfrac{1}{2}f(n) + f'(n)/12. \end{aligned}$$

If we aim to express this in terms of $F(x)$, the continuous cumulative distribution function corresponding to density $f(x)$, we use argument $n + h_n$; then

$$F(n + h_n) \doteq F(n) + h_n f(n) + \tfrac{1}{2}h_n^2 f'(n),$$

so that the choice of $h_n = \frac{1}{2}$ gives the correct coefficient to $f(n)$ and approximately the correct coefficient to $f'(n)$.

A more elaborate discussion links the cumulants of a continuous density and those of a discrete distribution formed from it by grouping, these relations forming Sheppard's corrections for grouping (Kendall and Stuart, 1969).

3.5 Inversion of series

In the previous sections we have discussed asymptotic expansions for integrals and, more briefly, for series but there are other important contexts in which asymptotic expansions arise and one such is in the solution of equations in which one or more of the terms has an asymptotic expansion.

In one simple version of this we are interested in the equation

$$g_n(x) = a, \tag{3.47}$$

where $g_n(x)$ has an asymptotic expansion, say, in powers of $1/\sqrt{n}$ as $n \to \infty$ with leading term $g^{(0)}(x)$ and that the solution of $g^{(0)}(x) = a$ is known. Without loss of generality we may take $a = 0$, $g^{(0)}(0) = 0$.

We thus have to solve

$$g^{(0)}(x) + g^{(1)}(x)/\sqrt{n} + g^{(2)}(x)/n + \cdots = 0 \tag{3.48}$$

for a solution near $x = 0$ and so we write

$$x = x^{(1)}/\sqrt{n} + x^{(2)}/n + \cdots$$

substitute in (3.48), expand by Taylor's theorem and equate coefficients of successive powers of $1/\sqrt{n}$.

We shall illustrate this technique in Chapter 4 in connexion with the calculation of the quantiles of distributions. A rigorous justification can often be obtained via Lagrange's theorem; see de Bruijn (1958).

We give here just one example, in which the functional form is simpler than in the general formulation (3.47).

Example 3.13 Quantiles of a special gamma distribution

Suppose that we wish to approximate the quantiles of the gamma distribution with density xe^{-x}, i.e. with distribution function $1-(x+1)e^{-x}$. To obtain the lower quantiles we must solve for small ε the equation

$$1 - e^{-x}(x+1) = \varepsilon, \qquad (3.49)$$

whereas to obtain the upper quantiles we solve for small η

$$e^{-x}(x+1) = \eta. \qquad (3.50)$$

Now the solution of (3.49) is small and expansion of the left-hand side shows that to a first approximation $\frac{1}{2}x^2 = \varepsilon$. This suggests writing $x_0 = \sqrt{(2\varepsilon)}$ and expanding (3.49) in the form

$$[2\{1 - e^{-x}(x+1)\}]^{1/2} = x - \frac{x^2}{3} + \tfrac{5}{72}x^3 + \cdots.$$

On equating this to x_0 and inverting, we have that

$$x = x_0 + \tfrac{1}{3}x_0^2 + \tfrac{11}{72}x_0^3 + \cdots. \qquad (3.51)$$

Effective approximation to the upper quantiles via (3.50) is more difficult. For small η, x will be large and the exponential term thus dominant. That suggests rewriting the equation as

$$x = l_\eta + \log(x+1),$$

where $l_\eta = -\log\eta$ and then defining a convergent iterative scheme via

$$x_{(k)} = l_\eta + \log(x_{(k-1)} + 1), \qquad x_{(1)} = l_\eta,$$

leading thereby to

$$x_{(2)} = l_\eta + \log(l_\eta + 1),$$
$$x_{(3)} = l_\eta + \log\{l_\eta + \log(l_\eta + 1)\}. \qquad (3.52)$$

Table 3.3 illustrates (3.51) and (3.52).

Table 3.3 *Expansion for quantiles of gamma distribution of index one.*

Lower quantile					*Upper quantile*				
ε	*exact*	*1 term*	*2 terms*	*3 terms*	η	*exact*	*1 term*	*2 terms*	*3 terms*
10^{-2}	0·1486	0·1414	0·1481	0·1485	10^{-2}	6·6384	4·6052	6·3289	6·5970
10^{-1}	0·5318	0·4472	0·5139	0·5275	10^{-1}	3·8897	2·3026	3·4973	3·8061
0·2	0·8244	0·6325	0·7658	0·8044	0·2	2·9943	1·6094	2·5686	2·8816
0·5	1·6783	1	1·3333	1·4861	0·5	1·6783	0·6931	1·2198	1·4905

3.6 Asymptotic expansions of distributions by direct methods

The central object of this book is the study of asymptotic expansions for probability distributions. The most obvious route is via an explicit expansion of the cumulative distribution function or other direct specification of the distribution, e.g. the probability density function. Usually, although not necessarily, we standardize the random variable first, before making the expansion, so that the leading term of the expansion corresponds for example to a normal distribution of fixed mean and variance.

In section 3.7 we consider what will be called indirect methods in which the expansion is made first via the moment generating function or characteristic function, and in section 3.8 we examine the relation with expansions of the random variables themselves.

Typically, although not necessarily, the expansion corresponds to large sample sizes or amounts of information. As always, the limiting operations do not have direct physical significance but are devices for producing useful approximations.

We shall discuss a number of examples.

Example 3.14 Student's t distribution

A simple instance of a direct expansion is provided by Student's t distribution with n degrees of freedom for which the probability density function is

$$p_n(t) = \frac{\Gamma(\frac{1}{2}n + \frac{1}{2})}{\sqrt{(n\pi)}\Gamma(\frac{1}{2}n)}(1 + t^2/n)^{-1/2n - 1/2}. \tag{3.53}$$

The random variable is already in standardized form so that we can work directly with the density and find by elementary expansion that

$$\log p_n(t) = -\tfrac{1}{2}\log(2\pi) - \tfrac{1}{2}t^2 + n^{-1}(t^4/4 - t^2/2 - 1/4) + O(n^{-2}),$$

where to compute the constant term Stirling's formula has been used.

It follows that

$$p_n(t) = \phi(t)\{1 + n^{-1}(t^4/4 - t^2/2 - 1/4)\} + O(n^{-2}),$$

where $\phi(t)$ is the standardized normal density function.

An alternative expression which provides a useful link with the discussion of Edgeworth expansions in Chapter 4 is obtained by expressing the correction term as a combination of Hermite polynomials so that the approximation becomes

$$\phi(t)[1 + n^{-1}\{H_2(t) + H_4(t)/4\} + O(n^{-2})]. \tag{3.54}$$

We shall see later that the term in $H_2(t)$ arises because the variance of Student's t is not exactly one and the term in $H_4(t)$ stems from the nonzero fourth cumulant.

An advantage of the form (3.54) is that it is easily formally integrated to give an approximation to the cumulative distribution function.

Example 3.15 Poisson distribution

Another familiar application is provided by the Poisson distribution regarded as the limit of a binomial distribution. For simplicity we again take the leading terms and just one correction term. If in the binomial distribution with n trials and probability of success θ, we set $n\theta = \mu$ and let $n \to \infty$, $\theta \to 0$ with μ fixed, we have for the probability π_r of r successes with r fixed:

$$\begin{aligned}\log \pi_r &= \log(\mu^r/r!) + (n-r)\log(1-\mu/n) + \log(1-1/n) + \cdots \\ &\quad + \log\{1-(r-1)/n\} \\ &= \log(\mu^r/r!) - \mu + n^{-1}(r\mu - \tfrac{1}{2}\mu^2 - \tfrac{1}{2}r^2 + \tfrac{1}{2}r) + O(n^{-2}),\end{aligned}$$

so that

$$\pi_r = \frac{e^{-\mu}\mu^r}{r!}\{1 + (r\mu - \tfrac{1}{2}\mu^2 - \tfrac{1}{2}r^2 + \tfrac{1}{2}r)/n + O(n^{-2})\}.$$

In particular, interest is often focused on π_0, for which we have that

$$\pi_0 = e^{-\mu}\{1 - \tfrac{1}{2}\mu^2/n + O(n^{-2})\},$$

illustrating a tendency for the Poisson distribution to overestimate the probability of zero; note the related fact that for a given mean the Poisson distribution has greater variance than the corresponding binomial distribution.

Other standard distributions can be studied in a similar fashion.

3.7 Asymptotic expansions of distributions via generating functions

We now consider expansions derived by first obtaining an expansion for the moment generating function or characteristic function and then inverting term by term; of course the legitimacy of this depends on some general theory. Much the most important expansions of this type are connected with the central limit theorem and it is convenient to discuss these in a separate chapter. Here we give some different and rather more specialized examples.

Example 3.16 Noncentral chi-squared

The noncentral chi-squared distribution with r degrees of freedom and noncentrality parameter Δ^2 has moment generating function

$$M_{r\Delta}(t) = (1-2t)^{-1/2r} \exp\{t\Delta^2(1-2t)^{-1}\}. \tag{3.55}$$

We consider the behaviour as $\Delta \to 0$. Because at $\Delta = 0$ the distribution reduces to the central chi-squared distribution no special standardization is needed. By direct expansion we have that

$$M_{r\Delta}(t) = (1-2t)^{-1/2r}\{1 + t\Delta^2(1-2t)^{-1} + \tfrac{1}{2}t^2\Delta^4(1-2t)^{-2} + O(\Delta^6)\}. \tag{3.56}$$

To achieve simple term-by-term inversion we express (3.56) in terms of the moment generating function of central chi-squared distributions by writing

$$t/(1-2t) = -\tfrac{1}{2} + \tfrac{1}{2}/(1-2t),$$
$$t^2/(1-2t)^2 = \tfrac{1}{4} - \tfrac{1}{2}/(1-2t) + \tfrac{1}{4}/(1-2t)^2.$$

Thus if $q_s(x)$ denotes the density of central chi-squared with s degrees of freedom, we have on inversion of (3.56) that $f_{r\Delta}(x)$, corresponding to $M_{r\Delta}(t)$, is such that as $\Delta \to 0$,

$$f_{r\Delta}(x) = q_r(x) + \tfrac{1}{2}\Delta^2\{q_r(x) - q_{r+2}(x)\}$$
$$+ \tfrac{1}{8}\Delta^4\{q_r(x) - 2q_{r+2}(x) + q_{r+4}(x)\} + O(\Delta^6). \tag{3.57}$$

Integration yields an exactly corresponding expansion for the cumulative distribution function. It is possible to express (3.57) in the alternative form

$$f_{r\Delta}(x) = q_r(x)\{1 + \tfrac{1}{2}\Delta^2 c_1(x) + \tfrac{1}{8}\Delta^4 c_2(x)\} + O(\Delta^6), \qquad (3.58)$$

where $c_s(x)$ is a polynomial in x with unit constant term, or equivalently as an expansion in generalized Laguerre polynomials. This form, while establishing a general relation with other expansions, is for computational purposes less useful than (3.57), especially in its integrated form.

A further possibility is to absorb the second term of (3.57) into the leading term by a rescaling of the random variable.

Now the mean and variance of the distribution (3.55) are respectively

$$r + \Delta^2,\ 2(r + 2\Delta^2). \qquad (3.59)$$

If we wish to examine other limiting forms via expansions in which (a) $r \to \infty$ for fixed Δ^2; (b) $\Delta^2 \to \infty$ for fixed r; (c) $r \to \infty$ with $\Delta^2 = r\kappa^2$, for fixed κ^2, then (3.59) indicates the appropriate standardization.

Example 3.17 Negative binomial and log series distributions

We approach the negative binomial distribution via its interpretation as a compound Poisson distribution. It is slightly more convenient to work with a probability generating function rather than with a moment generating function.

If in a Poisson distribution of mean μ, the value of μ is a random variable, the probability generating function of the resulting random variable X is

$$G(X; z) = E(z^X) = E(e^{\mu z - \mu}) = M(\mu; z - 1), \qquad (3.60)$$

where the second expectation is with respect to μ and where the moment generating function is that of μ. If now μ has a gamma distribution of index β and scale parameter θ, we have that

$$M(\mu; t) = (1 - \theta t)^{-\beta},$$

so that

$$G(X; z) = \{1 - (z - 1)\theta\}^{-\beta}. \qquad (3.61)$$

Various limiting forms can now be explored, depending on the relation between θ and β. One important case is based on the zero-truncated form of (3.61) considered for fixed θ as $\beta \to 0$, the limiting case corresponding to very large dispersion for μ. Now the probability generating function of the truncated distribution, i.e. of the conditional distribution given $X \neq 0$, is

$$G_c(X;z) = [\{1-(z-1)\theta\}^{-\beta} - (1+\theta)^{-\beta}]\{1-(1+\theta)^{-\beta}\}^{-1}.$$

On writing expressions of the form $\alpha^{-\beta}$ as $\exp(-\beta \log \alpha)$, we have as $\beta \to 0$ for fixed θ that

$$\begin{aligned} G_c(X;z) = 1 &- \log(1+\theta-\theta z)/\log(1+\theta) \\ &+ \tfrac{1}{2}\beta[\{\log(1+\theta-\theta z)\}^2/\log(1+\theta) \\ &+ \log(1+\theta-\theta z) - \log(1+\theta)] + O(\beta^2). \qquad (3.62) \end{aligned}$$

The leading term in this expansion defines the log series distribution. Of course it would be possible to carry out the argument directly in terms of the probabilities rather than via the probability generating function.

As we have pointed out a number of times, the parameter in terms of which expansions are expressed is often, but not invariably, a sample size or amount of information. In the last two examples the parameters Δ, β do not naturally have such an interpretation; in this broader context, the technique is one of perturbation theory. The general approach, which is of wide applicability especially in applied probability, is that of studying small variations from a problem with a known and preferably fairly simple solution. For further illustration, see Exercise 3.11.

3.8 Asymptotic expansions for random variables

The definitions of section 3.1 and the subsequent discussion all concern the expansion of functions defined by series or integrals or via explicit or implicit analytical expressions. As such, although all our examples have had some probabilistic motivation they have not been probabilistic in character.

We now consider expansions defined directly in terms of random variables. Suppose that $\{b_{hn}\}$ are a base set of sequences in the sense of section 3.1 and suppose that $\{Y_n\}$ is a sequence of continuous random variables such that

$$Y_n = X_0 + X_1 b_{1n} + \cdots + X_h b_{hn} + O_p(b_{h+1,n}), \qquad (3.63)$$

where $\{X_0, X_1, \ldots\}$ have a distribution not depending on n. It is natural to call (3.63) a stochastic asymptotic expansion; note that Y_n converges in distribution to X_0. We are typically interested in $h=0,1,2$, and very often in the special cases where $b_{1n}=1/\sqrt{n}$, $b_{2n}=1/n$, or where $b_{1n}=1/n$, $b_{2n}=1/n^2$, etc.

An important question concerns the relation between the expansion (3.63) and an asymptotic expansion of the corresponding distribution or probability density function.

To study this we concentrate on $h=1,2$ and $b_{1n}=1/\sqrt{n}$, $b_{2n}=1/n$, although all we really need is $b_{2n}=b_{1n}^2$. Suppose first that (3.63) holds and denote the joint density of (X_0, X_1, X_2) by $f_{012}(x_0, x_1, x_2)$, which is by definition independent of n. We denote the joint density of (X_0, X_1) similarly by $f_{01}(x_0, x_1)$. Then taking the case $h=1$ first we have formally that the probability density of Y_n is, with error $O(n^{-1})$,

$$\begin{aligned} f(Y_n; y) &= \int_{-\infty}^{\infty} f_{01}(y - x_1/\sqrt{n}, x_1)\,dx_1 \\ &= \int_{-\infty}^{\infty} \{f_{01}(y, x_1) - (x_1/\sqrt{n})\partial f_{01}(y, x_1)/\partial y\}\,dx_1 \\ &= f_0(y) - \int_{-\infty}^{\infty} x_1 \partial f_{01}(y, x_1)/\partial y\, dx_1/\sqrt{n}. \end{aligned} \tag{3.64}$$

Now the conditional mean of X_1 given $X_0 = y$ is

$$\mu_{1\cdot 0}(y) = \int_{-\infty}^{\infty} x_1 \{f_{01}(y, x_1)/f_0(y)\}\,dx_1,$$

so that we can rewrite (3.64) as

$$f(Y_n; y) = f_0(y) - \partial\{f_0(y)\mu_{1\cdot 0}(y)\}/\partial y/\sqrt{n} + O(n^{-1}). \tag{3.65}$$

If we retain a further term we have similarly that the remainder term in (3.65) can be written

$$n^{-1}[-\partial\{f_0(y)\mu_{01\cdot 0}(y)\}/\partial y + \tfrac{1}{2}\partial^2\{f_0(y)\mu_{2\cdot 0}(y)\}/\partial y^2] + O(n^{-3/2}), \tag{3.66}$$

where

$$\mu_{01\cdot 0}(y) = E(X_2 | X_0 = y), \qquad \mu_{2\cdot 0}(y) = E(X_1^2 | X_0 = y).$$

It may sometimes be more convenient to write (3.65) and (3.66) in terms of the cumulative distribution function of Y_n. On integration we obtain

$$F(Y_n; y) = F_0(y) - f_0(y)\mu_{1\cdot 0}(y)/\sqrt{n} - f_0(y)\mu_{01\cdot 0}(y)/n \\ + \tfrac{1}{2}\partial\{f_0(y)\mu_{2\cdot 0}(y)\}/\partial y/n + O(n^{-3/2}). \tag{3.67}$$

It is possible to amalgamate terms by amending the argument of the function $F_0(y)$ to, say,

$$y + \alpha_1(y)/\sqrt{n} + \alpha_2(y)/n$$

and choosing α_1 and α_2 to achieve agreement after expansion by Taylor's theorem. The result is to replace the right-hand side of (3.67) by F_0 with argument

$$y - \mu_{1\cdot 0}(y)/\sqrt{n} + \Bigg[-\mu_{01\cdot 0}(y) + \tfrac{1}{2}\mu'_{2\cdot 0}(y) \\ + \frac{f'_0(y)}{2f_0(y)}\{\mu_{2\cdot 0}(y) - \mu^2_{1\cdot 0}(y)\}\Bigg]n^{-1} + O(n^{-3/2}). \tag{3.68}$$

If and only if the modified argument (3.68) is a linear function of y we can regard the new distribution as corresponding to a scale and location change of the original distribution $F_0(y)$.

Example 3.18 Noncentral chi-squared for small noncentrality

As a simple example consider a noncentral chi-squared random variable W_m with m degrees of freedom and noncentrality parameter Δ. If $Z_1, \ldots, Z_m$ are independent and identically distributed random variables with standard normal distributions, we can write

$$W_m = Z_1^2 + \cdots + Z_m^2 + 2\Delta Z_1 + \Delta^2. \tag{3.69}$$

We suppose Δ to be small. To establish an exact parallel with the previous discussion we would write $\Delta = \phi/n$, but this is hardly necessary. Here, in the notation of the general discussion, X_2 is a constant, Δ^2, whereas the conditional mean of $X_1 = Z_1$ is zero by symmetry and the conditional mean square, given $X_0 = Z_1^2 + \cdots + Z_m^2 = y$, is y/m. It follows on substituting in (3.68) and using the special form for $f_0(y)$, the central chi-squared distribution with m degrees of freedom, that the cumulative distribution function of W_m is

$$F_0\{y(1 - \Delta^2/m)\} + O(\Delta^{3/2}).$$

Note that $E(W_m) = m(1 + \Delta^2/m)$, so that if we used a central chi-squared approximation with adjusted mean the approximation would be $F_0\{y(1+\Delta^2/m)^{-1}\}$. Sincc the support of W_m is the positive line it is natural to think of rescaled central chi-squared as a simple working approximation to noncentral chi-squared but the approximation is in a sense much better than might be expected and this is confirmed by numerical work.

Now if $f_0(y)\mu_{1\cdot 0}(y)$ is constant and in particular if $\mu_{1\cdot 0}=0$, then

$$f(Y_n;y)=f_0(y)+O(n^{-1}), \tag{3.70}$$

even though

$$Y_n = X_0 + O_p(n^{-1/2}).$$

We shall examine some applications of this in Chapter 4.

Now consider the converse problem in which an asymptotic expansion is given for the distribution function or density function and a possible expansion for a random variable is of interest. Suppose then that

$$f(Y_n;y)=f_0(y)\{1+a_1(y)/\sqrt{n}+a_2(y)/n\}+O(n^{-3/2}), \tag{3.71}$$

where $f_0(y)$ is a proper density function. Let X_0 be a random variable having the density $f_0(y)$. To match (3.71) with (3.65) and (3.66) we need only that

$$\begin{aligned} -\partial\{f_0(y)\mu_{1\cdot 0}(y)\}/\partial y &= a_1(y)f_0(y),\\ -\partial\{f_0(y)\mu_{01\cdot 0}(y)\}/\partial y+\tfrac{1}{2}\partial^2\{f_0(y)\mu_{21\cdot 0}(y)\}/\partial y^2 &= a_2(y)f_0(y). \end{aligned} \tag{3.72}$$

These equations can be satisfied in a great many ways, but in particular we can take X_1, X_2 to be sure functions of X_0, $X_1=b_1(X_0)$, $X_2=b_2(X_0)$, say. We then require that

$$\begin{aligned} &-\partial\{f_0(y)b_1(y)\}/\partial y = a_1(y)f_0(y),\\ &-\partial\{f_0(y)b_2(y)\}/\partial y = a_2(y)f_0(y)-\tfrac{1}{2}\partial^2\{f_0(y)b_1^2(y)\}/\partial y^2, \end{aligned}$$

which determine b_1, b_2 via for example

$$b_1(y)=-\{f_0(y)\}^{-1}\int_{-\infty}^{y} a_1(z)f_0(z)dz. \tag{3.73}$$

That is, we can, in particular, write in view of the expansion for the density a corresponding expansion for the random variable as

$$Y_n = X_0 + b_1(X_0)/\sqrt{n} + b_2(X_0)/n + O_p(n^{-3/2}). \tag{3.74}$$

The main applications of this will appear in Chapter 4 in connexion with expansions connected with the normal distribution.

3.9 Asymptotic expansions with a second parameter

It frequently happens that the quantity, Q say, for which an expansion is required depends on some extra parameter γ and that the character of the expansion changes drastically if γ passes from one region of its possible values to another. There are many discussions of this type of problem scattered over the literature and the field is somewhat marked by *ad hoc* methods. We do not aim to give anything like a comprehensive treatment here but will restrict our discussion to indicating a procedure which in a number of important cases leads to a rather elegant solution. We modify the original expansion into one whose leading term is expressed in terms of γ. To show how this may be done we consider a case of particular interest in statistical theory.

Let

$$Q_\tau(\gamma) = \int_{-\infty}^{\gamma} q(x)\phi(x; \tau^{-1})dx, \tag{3.75}$$

where we assume that $q(0) \neq 0$ and where, as usual,

$$\phi(x; \kappa) = \frac{1}{\sqrt{(2\pi)}\sqrt{\kappa}} e^{-1/2\kappa^{-1}x^2},$$

and consider the asymptotic behaviour of $Q_\tau(x)$ as $\tau \to +\infty$. By Laplace's method,

$$Q_\tau(\lambda) = \begin{cases} |\gamma|^{-1}q(\gamma)\tau^{-1}\phi(\gamma; \tau^{-1})\{1 + O(\tau^{-1})\} & \text{if } \gamma < 0, \\ \tfrac{1}{2}q(0)\{1 + O(\tau^{-1/2})\} & \text{if } \gamma = 0, \\ q(0)\{1 + O(\tau^{-1})\} & \text{if } \gamma > 0. \end{cases} \tag{3.76}$$

Note the substantial differences between the three cases, in regard both to the leading terms and the orders of the error terms. The differences occur, in essence, because the end point of integration coalesces with the maximum of ϕ over the region of integration as γ passes from the interval $(-\infty, 0)$ into $[0, \infty)$.

If q was constant, $q(x) = c$, we would of course have, exactly, $Q_\tau(\gamma) = c\Phi(\gamma\sqrt{\tau})$ and we now employ this fact to rewrite Q_τ by partial integration, as

$$Q_\tau(\gamma) = q(0)\Phi(\gamma\sqrt{\tau}) + \int_{-\infty}^{\gamma} \{q(x) - q(0)\}\phi(x; \tau^{-1})dx$$

$$= Q_\tau(\infty)\Phi(\gamma\sqrt{\tau}) - \tau^{-1}\phi(\gamma; \tau^{-1})\frac{q(\gamma) - q(0)}{\gamma}$$

$$+ \tau^{-1}\int_{-\infty}^{\gamma} k'(x)\phi(x; \tau^{-1})dx + O(\tau^{-1}), \tag{3.77}$$

where k' is the derivative of $k(x) = \{q(x) - q(0)\}/x$.

It follows from (3.77) that, under mild regularity of the function q,

$$Q_\tau(\gamma) = Q_\tau(\infty)\Phi(\gamma\sqrt{\tau}) - \frac{q(\gamma) - q(0)}{\gamma}\tau^{-1}\phi(\gamma; \tau^{-1})\{1 + O(\tau^{-1})\} \tag{3.78}$$

uniformly for $\gamma \in R$. The same result holds if in (3.75) we replace q by a function q_τ possessing an asymptotic expansion in powers of τ^{-1} with leading terms $q(x)$, in which case

$$Q_\tau(\gamma) = \int_{-\infty}^{\gamma} q_\tau(x)\phi(x; \tau^{-1})dx. \tag{3.79}$$

The precise regularity conditions for (3.78) and the extension to higher-order expansions are given by Temme (1982).

Often a preliminary transformation is needed to bring a given integral to the form (3.79).

Example 3.19 Incomplete gamma function

The gamma distribution function

$$F_\tau(\alpha) = \frac{1}{\Gamma(\tau)}\int_0^{\alpha} y^{\tau-1}e^{-y}dy$$

may, by the transformation $x = r(y)/\tau$, where

$$r(y) = \operatorname{sgn}(y - 1)\{2(-\log y + y - 1)\}^{1/2},$$

be recast as (3.79) with $\gamma = r(\alpha/\tau)$ and

$$q_\tau(x) = \{\tilde{\Gamma}(\tau)/\Gamma(\tau)\}\{x/(y/\tau - 1)\},$$

where

$$\tilde{\Gamma}(\tau) = (2\pi)^{1/2}\tau^{\tau-1/2}e^{-\tau}$$

is Stirling's approximation to $\Gamma(\tau)$. Hence, with ζ defined by $r(\zeta) = \gamma$, we have

$$F_\tau(\alpha) = \Phi(\sqrt{\tau}\gamma) + \{\gamma^{-1} - (\zeta - 1)^{-1}\}\tau^{-1}\phi(\gamma; \tau^{-1})\{1 + O(\tau^{-1})\},$$

uniformly in α.

Further results and exercises

3.1 Dawson's function is defined by

$$D(x) = e^{-(1/2)x^2}\int_0^x e^{1/2t^2}\,dt.$$

By writing $x - t = v$ and expanding about $v = 0$, show that for large x

$$D(x) = 1/x + 1/x^3 + \cdots.$$

A time-varying Poisson process with rate $\exp(\alpha + \beta t + \gamma t^2)$ is observed for $(0, t_0)$ and points are observed to occur at $x_1, \ldots, x_n$. Show that the log likelihood is

$$\alpha + \beta\sum x_j + \gamma\sum x_j^2 - \int_0^{t_0} \exp(\alpha + \beta t + \gamma t^2)dt,$$

so that it can be expressed in terms of the standardized normal integral ($\gamma < 0$) and Dawson's function ($\gamma > 0$). How does the information matrix for (α, β, γ) behave as $t_0 \to \infty$ in the three cases $\gamma < 0$, $\gamma = 0$, $\gamma > 0$?

[Sections 3.2, 3.3]

3.2 Show that under mild regularity conditions a valid asymptotic expansion for the integral

$$I_\tau(f) = \int_{-\infty}^{+\infty} f(t)\phi(t; \tau^{-1})dt \tag{3.80}$$

is given by

$$I_\tau(f) \sim \sum_{\nu=0}^{\infty} f^{(2\nu)}(0)2^\nu \frac{\Gamma(\nu + \frac{1}{2})}{\Gamma(\frac{1}{2})}\tau^{-\nu}, \tag{3.81}$$

for $\tau \to \infty$. In (3.80) f denotes a function on the real line which possesses a Taylor expansion around 0 and $\phi(t; \tau^{-1})$ is the probability density function of the normal distribution with mean 0 and variance τ^{-1}.

[Section 3.3]

3.3 Apply Laplace's method to show that as $n \to \infty$, for $a_0, a_1 > 0$, $ca_1 > a_0$,

$$\int_{-\infty}^{\infty} \frac{e^{na_0 x}}{(1 + be^{na_1 x})^c} g(x)dx$$
$$= \frac{\sqrt{(2\pi)}}{n} \frac{a_0^{a_0/a_1 - 1/2}(ca_1 - a_0)^{c - a_0/a_1 - 1/2}}{c^{c-1/2} a_1^c b^{a_0/a_1}} g(\tilde{x})\{1 + O(n^{-1})\},$$

where $e^{na_1\tilde{x}} = a_0/\{b(ca_1 - a_0)\}$.

Discuss in more detail the special cases $a_0 = a_1$, $b = 1$, $c = 2$, $g(x) = x, x^2$, specifying the results in terms of the mean and variance of the logistic density.

[Section 3.3]

3.4 *Method of stationary phase.* Consider an integral of the form

$$w(n) = \int_a^b q(y)e^{inr(y)}dy$$

where q and r are real functions, and $r'(y)$ is assumed to be 0 at a single point $c \in (a, b)$. Suppose also that $r''(c) \neq 0$.

Show that, for $n \to \infty$,

$$w(n) = e^{\operatorname{sgn} r''(c) i\pi/4} \sqrt{(2\pi)}\{n|r''(c)|\}^{-1/2} q(c)e^{inr(c)} + O(n^{-1})$$

(cf., for instance, Bleistein and Handelsman, 1975, or Olver, 1974).

[Section 3.3]

3.5 In section 3.3 the asymptotic expansion of a Laplace transform was derived formally by term-by-term integration of a Taylor series. Justification of this and a much more general such expansion can be derived from Watson's lemma: let $g(t)$ have an asymptotic expansion as $t \to 0$ in the form

$$\sum g_s t^{(s + \lambda - \mu)/\mu}$$

where $\lambda, \mu > 0$. Then provided the Laplace transform (3.12) exists for all sufficiently large z it has an asymptotic expansion as the argument $z \to \infty$, namely

$$\sum \Gamma\{(s + \lambda)/\mu\} g_s z^{-(s+\lambda)/\mu}.$$

[Section 3.3; Olver, 1974, p. 71]

3.6 Especially in connexion with amalgamation of expansions into the leading term, it may be convenient to approximate a given function by a rational function. The ratio $a_p^{(q)}(x)/b_q^{(p)}(x)$, where $a_p^{(q)}(x)$ and $b_q^{(p)}(x)$ are polynomials of degrees p and q, with $b_q^{(p)}(0) = 1$, is called the (p, q) Padé approximant to a given function $f(x)$, assumed expandable in a Taylor series about $x = 0$, if as $x \to 0$,

$$a_p^{(q)}(x) - f(x)b_q^{(p)}(x) = O(x^m)$$

for maximum possible m; often but not always $m = p + q + 1$. A Padé table is the set of approximations $p, q = 0, 1, \dots$ arranged conventionally in columns according to p and in rows according to q. In some applications the zeros of the denominator (and conceivably also the numerator) are, however, fixed *a priori*. There is a close connexion with continued fractions, corresponding to an iterative determination of $a_p^{(q)}(x)$ and $b_q^{(p)}(x)$ via the coefficients of the expansion of $f(x)$. In particular show that part of the Padé table for e^x is

$1/1$	$(1+x)/1$	$(1+x+\frac{1}{2}x^2)/1$
$1(1-x)$	$(1+\frac{1}{2}x)/(1-\frac{1}{2}x)$	$(1+\frac{2}{3}x+\frac{1}{6}x^2)/(1-\frac{1}{3}x)$
$1/(1-x+\frac{1}{2}x^2)$	$(1+\frac{1}{3}x)/(1-\frac{2}{3}x+\frac{1}{6}x^2)$	$(1+\frac{1}{2}x+\frac{1}{12}x^2)/(1-\frac{1}{2}x+\frac{1}{12}x^2)$.

Phillips (1982) has used Padé approximants systematically in approximating probability distributions. A rather different use is to obtain improved numerical approximations to functions for which initially only a few terms of a Taylor expansion are available. For an application to the virial expansion, partition functions and likelihood for point processes, see Ogata and Tanemura (1984).

[Section 3.3; Baker and Graves-Morris, 1981]

3.7 If $\{a_j\}$ and $\{b_j\}$ are two sequences, $A_j = a_1 + \cdots + a_j$, prove that

$$\sum a_j b_j = \sum A_j(b_j - b_{j+1}) + A_{n-1} b_n,$$

a formula sometimes called summation by parts. Discuss why this is likely to be useful if $\{a_j\}$ is rapidly oscillating and $\{b_j\}$ slowly varying, examining in particular the special case $a_j = (-1)^j$. Suppose that b_j is the probability in some discrete distribution and that $a_j = \cos(\omega j)$ and that $n \to \infty$. Discuss the relation of the resulting formula with the properties of characteristic functions.

[Section 3.4]

3.8 Suppose that a function $f(z)$ can be expanded in a series of inverse powers first of z and then of $z-a$, for suitable a. That is,

$$f(z)=\sum_{j=0}^{\infty} b_j/z^{j+1}=\sum_{j=0}^{\infty} c_j/(z-a)^{j+1}.$$

Prove formally by writing $z^{-j-1}=(z-a)^{-j-1}\{1+a/(z-a)\}^{-j-1}$, expanding by the binomial theorem and equating coefficients, that

$$c_j=a^j[\Delta^j(b_m a^{-m})]_{m=0},$$

where Δ is the forward difference operator. Put $a=1$, $z=-1$ to obtain Euler's transformation

$$\sum_{j=0}^{\infty}(-1)^j b_j=\sum_{j=0}^{\infty}(-1)^j\Delta^j b_0/2^{j+1}.$$

Obtain an approximation for the probability that a Poisson distributed random variable of mean μ is odd and make some numerical comparisons.

[Section 3.4]

3.9 The infinite integral

$$I=\int_{-\infty}^{\infty} g(x)dx$$

is to be estimated by the infinite sum

$$S(t,h)=h\sum_{j=-\infty}^{\infty} g(t+jh),$$

where t is an arbitrary origin. Suppose that t is chosen at random, being a uniformly distributed random variable, T, over $(0,h)$. Show that

$$E\{S(T,h)\}=I,$$

$$E\{S(T,h)\}^2=h\sum_{j=-\infty}^{\infty}\int_{-\infty}^{\infty} g(t)g(t+jh)dt.$$

Note that $g(t)$ is periodic and thus show via a Fourier series representation combined with Parseval's theorem that

$$\operatorname{var}\{S(T,h)\}=2\pi\sum_{j=-\infty}^{\infty}{}' |\tilde{g}(2\pi j/h)|^2,$$

where the summation excludes $j=0$ and where $\tilde{g}$ is the Fourier transform

$$\tilde{g}(p) = \frac{1}{\sqrt{(2\pi)}} \int_{-\infty}^{\infty} e^{ipx} g(x)\,dx.$$

Hence show that if $g(x) = \phi(x)$, the standard normal integral, $\text{var}\{S(T, h)\}$ can be expanded in terms of theta functions,

$$\text{var}\{S(T, h)\} = 2 \sum_{j=1}^{\infty} \exp\left(-\frac{4j^2\pi^2}{h^2} \right),$$

so that even for h as large as 2 or 3, the variance is given closely by the first term and is small.

[Section 3.4; Yates, 1948; Kendall, 1942; Moran, 1950]

3.10 Develop an expansion parallel to that of Example 3.13 but for the upper and lower tail of the distribution with density $\phi e^{-x} + k(1-\phi)e^{-kx}$, where $k \gg 1$.

[Section 3.5]

3.11 Consider a controlled pure birth process in which there is a target unit growth rate; if there are N_t individuals at time t, the current birth rate is assumed to be $g(N_t - t)$ where $g(0) = 1$, $g'(0) < 0$. Show that if $p(x, t)$ is the probability that there are $x + t$ individuals alive at time t, then

$$\frac{\partial p(x,t)}{\partial t} = \frac{\partial p(x,t)}{\partial x} - g(x)p(x,t) + g(x-1,t),$$

subject to randomization of the initial conditions to destroy periodicity in $N_t - t$. Suppose that we write $g(x) = g_\lambda(x) = h(x/\lambda)$, where $h(0) = 1$, $h'(0) < 0$ and let $\lambda \to \infty$ to represent a 'gentle' controlling mechanism. By introducing a new state variable $Z_t = X_t/\sqrt{\lambda}$ show that as $\lambda \to \infty$ the density of Z_t is asymptotically $(\pi a)^{-1/2} \exp(-z^2/a)$, where $a = \{-h'(0)\}^{-1}$.

[Section 3.7; Cox and Isham, 1978, 1980]

3.12 Let W_n have a chi-squared distribution with n degrees of freedom and take $Y_n = (W_n - n)/\sqrt{(2n)}$. Show that if X_0 has a standard normal distribution,

$$Y_n = X_0 + n^{-1/2}\sqrt{2}(X_0^2 - 1)/3 + O_p(n^{-1}),$$

and hence that

$$W_n^c = n^c[1 + n^{1/2}c\sqrt{2}X_0 + n^{-1}\{\tfrac{2}{3}c(X_0^2 - 1) + c(c-1)X_0^2\} \\ + O_p(n^{-3/2})]$$

and hence that if and only if $c = \frac{1}{3}$ the term in n^{-1} is constant. This gives the Wilson–Hilferty transformation of the chi-squared distribution to near normality.
[Section 3.8; Cox and Reid, 1987; Kendall and Stuart, 1969, p. 399]

3.13 Show that if $f(x) = O(x^a)$ for $x \to \infty$ and for some $a \geqslant 0$ then (under mild regularity conditions) as $\alpha \to \infty$

$$\frac{\alpha^\lambda}{\Gamma(\lambda)} \int_0^\infty f(x)x^{\lambda-1}e^{-\alpha x}dx \sim f(\mu) + O\left(\frac{\mu}{(1+\mu)^{2+a}}\alpha^{-1}\right)$$

uniformly with respect to $\mu \in (0, \infty)$, where $\mu = \lambda/\alpha$. Discuss higher-order expansions.

[Section 3.9; Temme, 1983]

3.14 Let

$$Q_\tau(\gamma) = \int_{-\infty}^{\gamma} \frac{1}{1+x^2}\phi(x; \tau^{-1})dx.$$

Give a probabilistic interpretation of this integral and show that

$$Q_\tau(\gamma) = \{1 - \Phi(\sqrt{\tau})\}\Phi(\sqrt{\tau}\gamma) + \frac{\gamma}{1+\gamma^2}\tau^{-1}\phi(\gamma; \tau^{-1})\{1 + O(\tau^{-1})\}$$

uniformly in γ.

[Section 3.9; Temme, 1982]

Bibliographic notes

The material in sections 3.1–3.5 is largely contained in the general literature on asymptotic expansions. Olver's (1974) comprehensive book has some emphasis on the special functions of mathematical physics, whereas de Bruijn (1958) has a very illuminating discussion of some of the more analytical issues.

There is a close connexion well short of total identity between the discussion of asymptotic expansions for special distributions and the derivation of simple numerical approximations for properties of those distributions; see Molenaar (1970) for an account of the latter. More broadly there is a loose connexion with approximation theory (Rice, 1964, 1969). In that theory a given function is to be approximated by a suitable combination of given functions, a defined measure of distance between true and approximate functions to be minimized. Here there is typically no notion of limiting form involved, but the general form of the approximation has to be specified *a priori*. Some combination of the methods could be used, for example, in contexts such as the modification of Stirling's formula.

For a multivariate version of the Euler–Maclaurin formula, see Bhattacharya and Rao (1976).

The continuity correction, motivated by geometrical considerations, has a long history in statistical contexts; see Cox (1970) for the rather more formal approach adopted here and Cameron (1986). The discussion of the relation between the stochastic expansion of random variables and that of distribution functions is based on Cox and Reid (1987).

Lindley (1961) applied Laplace's method to Bayesian calculations. The important development to the calculation of expectations sketched in Example 3.9 is due to Tierney and Kadane (1986).

For the relation between asymptotic expansions and nonstandard analysis, see van den Berg (1987).

For a rather comprehensive discussion of the technique of matched asymptotic expansions indicated in section 3.9, see Temme (1985). Van Dyke (1975, Chs 5 and 6) discusses the use of matched asymptotic expansions in fluid mechanics; Eckhaus (1973) gives a more mathematical discussion and Simmonds and Mann (1986) a clear introduction.

CHAPTER 4

Edgeworth and allied expansions

4.1 Introduction

In the previous chapters we have given both the general ideas about asymptotic expansions and a number of examples; we have, however, deliberately postponed the most important example for separate discussion. This example concerns the behaviour of sums of independent random variables often of identically distributed random variables. In special cases it is possible to find the convolution integral or sum explicitly and to derive any asymptotic expansion from that explicit form, for example by using the techniques of section 3.6. For a discussion of any generality we must, however, use generating functions.

In the present chapter we first derive the direct Edgeworth expansion and later its inversion, the Fisher–Cornish expansion. This inversion is closely related to a representation of the random variable concerned via a stochastic asymptotic expansion in terms of a standard normal variable. Also we discuss the saddlepoint expansion, or, as we shall call it, the tilted expansion, the leading term of which gives at least the same asymptotic accuracy as two terms of a direct Edgeworth expansion.

4.2 Direct Edgeworth expansion

Let $Y_1, \ldots, Y_n$ be independent and identically distributed random variables, copies of a random variable Y with $E(Y) = \mu$, $\text{var}(Y) = \sigma^2$ and with cumulants κ_r; we write $\rho_r = \kappa_r/\sigma^r$ for the standardized cumulants. As in section 1.2 we have that

$$S_n = Y_1 + \cdots + Y_n, \qquad S_n^* = (S_n - n\mu)/(\sigma\sqrt{n}),$$

so that the cumulant generating function of S_n^* is

$$K(S_n^*; t) = -\sqrt{n}\mu t/\sigma + nK\{Y; t/(\sigma\sqrt{n})\}. \tag{4.1}$$

As noted in Chapter 1 there are some advantages, especially in the multivariate case, in not dividing by σ, but for the present purpose we shall include that factor.

Now the standard derivation of the central limit theorem is via the evaluation of the limit of (4.1) as $n \to \infty$, namely $\frac{1}{2}t^2$. We now develop the argument further by first writing

$$K(Y;t) = \mu t + \tfrac{1}{2}\sigma^2 t^2 + \tfrac{1}{6}\rho_3\sigma^3 t^3 + \tfrac{1}{24}\rho_4\sigma^4 t^4 + \cdots$$

and then substituting into (4.1). Expansion then gives that

$$K(S_n^*;t) = \tfrac{1}{2}t^2 + \rho_3 t^3/(6\sqrt{n}) + \rho_4 t^4/(24n) + O(n^{-3/2}). \qquad (4.2)$$

The moment generating function is thus given on taking exponentials as

$$M(S_n^*;t) = \exp(\tfrac{1}{2}t^2)\{1 + \rho_3 t^3/(6\sqrt{n}) + \rho_4 t^4/(24n) + \rho_3^2 t^6/(72n) + O(n^{-3/2})\}. \qquad (4.3)$$

This has to be inverted to obtain the probability density or distribution function. For this we use the result that

$$\int e^{tx}\phi(x)H_r(x)dx = t^r \exp(\tfrac{1}{2}t^2), \qquad (4.4)$$

where $H_r(x)$ is the Hermite polynomial of degree r as defined in section 1.6 via

$$(d/dx)^r \phi(x) = (-1)^r H_r(x)\phi(x), \qquad (4.5)$$

with $\phi(x) = \exp(-\frac{1}{2}x^2)/\sqrt{(2\pi)}$.

Formal inversion of (4.3) now yields the expansion

$$f(S_n^*;x) = \phi(x)\{1 + \rho_3 H_3(x)/(6\sqrt{n}) + \rho_4 H_4(x)/(24n) + \rho_3^2 H_6(x)/(72n)\} + O(n^{-3/2}) \qquad (4.6)$$

for the probability density function, and, on integration, using (4.5), the expansion

$$F(S_n^*;x) = \Phi(x) - \phi(x)\{\rho_3 H_2(x)/(6\sqrt{n}) + \rho_4 H_3(x)/(24n) + \rho_3^2 H_5(x)/(72n)\} + O(n^{-3/2}) \qquad (4.7)$$

for the distribution function.

Equations (4.6) and (4.7) form the Edgeworth expansions of S_n^*. Some immediate comments are as follows:

1. It would be possible to continue with further terms although we shall not have the occasion to do this;
2. It would be possible also to stop with the first correction term, having an error of order n^{-1} and this is indeed useful if the main aspect of nonnormality of concern is skewness;
3. The function

$$\phi(x)\{1 + \rho_3 H_3(x)/(6\sqrt{n})\} \tag{4.8}$$

formed from two terms of (4.6) is for fixed n and $\rho_3 \neq 0$ not a probability density in x. This is because for sufficiently large negative $x\rho_3$ the value of (4.8) is negative. This does not conflict with the asymptotic property for fixed x as $n \to \infty$. On the other hand, if the n^{-1} terms are included the resulting function is a probability density for $\rho_4/n \leqslant 2.2$ and for sufficiently small values of ρ_3^2/n;
4. So far as the density is concerned, the difficulty over nonpositivity can be circumvented by writing (4.8) in the form

$$\phi(x)\exp\{\rho_3 H_3(x)/(6\sqrt{n})\} \tag{4.9}$$

which of course agrees with (4.8) to the requisite order. This form has, however, the serious disadvantage of being unbounded and therefore certainly not being capable of exact normalization on $(-\infty, \infty)$. This may be overcome by restricting the distribution to some finite interval, or by inserting a convergence factor, for example replacing (4.9) by

$$\phi(x)\exp\{\rho_3 H_3(x)/(6\sqrt{n}) - \rho_3^2 H_3^2(x)/(72n)\}. \tag{4.10}$$

Both these devices involve, however, some substantial arbitrariness. There is the further disadvantage that explicit integration of (4.9) to achieve an expansion for the cumulative distribution function requires additional expansion;
5. We shall see in section 6.3 that if the expansion is to be based on a normal distribution of arbitrary variance the most natural approach is to use the one-dimensional form of the tensorial Hermite polynomials associated with that distribution, these being different from the standard Hermite polynomials except when the normal distribution is the standard one. Suppose, however, that an expansion in standard Hermite polynomials is to be carried out starting not from (4.2) but rather from a cumulant generating function of the general form

$$K(Z_n^*;t) = (\mu'/\sqrt{n} + \mu''/n)t + \tfrac{1}{2}(1 + \kappa'/\sqrt{n} + \kappa''/n)t^2$$
$$+ (\rho_3 + \rho_3'/\sqrt{n})t^3/(6\sqrt{n}) + \rho_4 t^4/(24n) + O(n^{-3/2}), \tag{4.11}$$

where μ', μ'', κ', κ'', ρ_3' are constants specifying departures from the simpler form treated earlier. Note that Z_n^* need not be a linear function of sums of independent and identically distributed random variables. Then on exponentiation we have that

$$M(Z_n^*;t) = e^{(1/2)t^2}\{1 + (\mu' t + \tfrac{1}{2}\kappa' t^2 + \rho_3 t^3/6)/\sqrt{n}$$
$$+ (\mu'' t + \tfrac{1}{2}t^2 + \rho_3' t^3/6 + \rho_4 t^4/24)/n + \tfrac{1}{2}(\mu' t + \tfrac{1}{2}\kappa' t^2$$
$$+ \rho_3 t^3/6)^2/n\} + O(n^{-3/2}). \tag{4.12}$$

This can now be inverted into an expansion for the probability density function $f(Z_n^*;z)$.

Note first the relatively complicated form of the n^{-1} term. If we take the expansion only to the term in $1/\sqrt{n}$ we have that

$$f(Z_n^*;z) = \phi(z)[1 + \{\mu' H_1(z) + \tfrac{1}{2}\kappa' H_2(z) + \rho_3 H_3(z)/6\}n^{-1/2}]$$
$$+ O(n^{-1}). \tag{4.13}$$

Of course if we considered instead of Z_n^* a new random variable

$$Z_n^{*\prime} = (Z_n^* - \mu'/\sqrt{n})/(1 + \kappa'/\sqrt{n})^{1/2}$$

the terms in $H_1(z)$ and $H_2(z)$ would disappear. See Example 4.4 for further illustration;

6. If the components Y_j are independent but not identically distributed, we have that

$$\kappa_r(S_n) = \sum \kappa_r(Y_j).$$

We standardize S_n in the usual way to

$$S_n^* = (S_n - \sum \mu_j)/\sqrt{\sum \sigma_j^2},$$

where

$$\mu_j = E(Y_j) = \kappa_1(Y_j), \qquad \sigma_j^2 = \text{var}(Y_j) = \kappa_2(Y_j)$$

and then formally the Edgeworth expansion is obtained, provided that

$$\sum \kappa_3(Y_j)/(\sum \sigma_j^2)^{3/2} \qquad \sum \kappa_4(Y_j)/(\sum \sigma_j^2)^2$$

are bounded as $n \to \infty$. The rate of decay is settled by the reciprocal of the total 'information', i.e. by the total variance $\sum \sigma_j^2$; typically

$\sum\sigma_j^2/n$ will tend to a positive limit as $n \to \infty$ and then the orders in (4.6) and (4.7) will be retained;

7. The discussion above applies to continuous random variables. If the Y_j are integer-valued random variables, so is S_n and hence S_n^* has atoms at a spacing of order $1/\sqrt{n}$, i.e. the distribution function of S_n^* is a step function with the indicated structure. The expansion (4.6) remains valid. It can be shown that this implies that while the $1/\sqrt{n}$ term in the expansion of the cumulative distribution function has the required properties, for the $1/n$ term to be meaningful it is necessary to evaluate it at the points of support and to apply a continuity correction. See Example 4.2 below for detailed illustration.

   In more complicated discrete cases it may be difficult to specify the points of support, often because these form a rather irregular set. It is then hard to give a discussion of any generality, although numerical work suggests that there will often be an improvement over simple use of a normal approximation. For an important family of rank statistics, a careful analysis (Does, 1983) shows that the Edgeworth expansion is valid for the cumulative distribution function up to and including the $1/n$ term, ultimately because the individual atoms of probability are sufficiently small.
8. A different formulation is obtained by looking at a local average, e.g. at

$$\int F_n(x - z)\varepsilon^{-1}w(z/\varepsilon)dz, \tag{4.14}$$

where $F_n(x)$ is the distribution function under approximation and

$$w(z) \geqslant 0, \qquad \int w(z)dz = 1 \tag{4.15}$$

and ε is a small quantity, regarded as fixed as $n \to \infty$. Then the $1/n$ term is applicable and a better approximation thereby obtained to a locally smoothed version of the function.

We now consider two examples where exact calculation is possible. This both enables numerical comparisons to be made between asymptotic series and the 'true' distribution and also allows direct derivation of the asymptotic expansion for comparison with the indirect derivation via the moment generating function.

Example 4.1 Gamma distribution

Let $Y_1, \ldots, Y_n$ be independent and identically distributed random variables, copies of a random variable Y having an exponential distribution of unit mean. Thus S_n has the density

$$z^{n-1}e^{-z}/(n-1)!$$

and $E(S_n) = n$, $\text{var}(S_n) = n$, $\rho_3 = 2$, $\rho_4 = 6$. The standardized sum is thus $S_n^* = (S_n - n)/\sqrt{n}$ and the Edgeworth expansion for its density is

$$\phi(x)\{1 + H_3(x)/(3\sqrt{n}) + H_4(x)/(4n) + H_6(x)/(18n)\} + O(n^{-3/2}). \tag{4.16}$$

Now the exact density of S_n^* is

$$\sqrt{n}(n + x\sqrt{n})^{n-1} \exp(-n - x\sqrt{n})/(n-1)! \tag{4.17}$$

and direct expansion, initially of the log density, including the use of Stirling's theorem, recovers (4.16) at the end of some detailed calculation.

Numerical work confirms the main points that are qualitatively obvious about the comparison of (4.16) and (4.17) and we shall not give extensive details. If the expansion is applied for small values of n, there is the major difficulty that the supports of exact and approximating distributions do not agree and we can therefore expect very poor performance in the lower tail. This is confirmed in Table 4.1, which shows for $n = 2, 5$, admittedly both very small values

Table 4.1 *Edgeworth approximation to the standardized gamma distribution, $n = 2, 5$. Standardized variable, x; originating variable, z.* Edg_0, *leading term, standard normal.* Edg_1, *with $1/\sqrt{n}$ correction term.*

			$n=2$			$n=5$	
x	Edg_0	z	*Exact*	Edg_1	z	*Exact*	Edg_1
−2·5	0·0175	−1·54	·	−0·0160	−0·59	·	−0·0037
−2·0	0·0540	−0·83	·	0·0285	0·53	0·0043	0·0379
−1.5	0·1295	−0·12	·	0·1639	1·65	0·1319	0·1512
−1·0	0·2420	0·59	0·4612	0·3560	2·76	0·3428	0·3141
−0·5	0·3521	1·29	0·5019	0·4662	3·88	0·4361	0·4242
0	0·3989	2·00	0·3828	0·3989	5·00	0·3924	0·3989
1	0·2420	3·41	0·1589	0·1279	7·24	0·1840	0·1698
2	0·0540	4·83	0·0546	0·0794	9·47	0·0577	0·0701
3	0·0044	6·25	0·0172	0·0232	11·71	0·0144	0·0163

of n, that while the introduction of the term in $1/\sqrt{n}$ brings about some improvement the expected discrepancies do occur. The situation is rather better in the upper tail.

Example 4.2 Poisson distribution

To illustrate performance on a discrete problem, we suppose that $Y_1, \ldots, Y_n$ are independent with Poisson distributions of unit mean so that S_n has a Poisson distribution of mean n. All the cumulants of the Poisson distribution are equal, so that $\rho_3 = \rho_4 = 1$. The standardized random variable is thus $S_n^* = (S_n - n)/\sqrt{n}$ and the formal expansion (4.7) for its cumulative distribution function is

$$\Phi(x) - \phi(x)\{H_2(x)/(6\sqrt{n}) + H_3(x)/(24n) + H_5(x)/(72n)\} + O(n^{-3/2}). \tag{4.18}$$

To use this to approximate to $P(S_n \leqslant r)$, for r not in the extreme tail of the distribution of S_n, we apply a continuity correction and take

$$x = (r - n + \tfrac{1}{2})/\sqrt{n},$$

the approximation to be considered only at integer values of r. Table 4.2 gives some numerical examples of this.

The inclusion of the $n^{-1/2}$ correction term substantially improves the approximation, with a much smaller overall improvement from the n^{-1} term. While there are better semi-empirical approximations to the cumulative distribution function, the Edgeworth approximation is remarkably good, even for $n = 4$.

Table 4.2 *Approximation to cumulative distribution function of Poisson distribution of mean n. Exact: $P(N \leqslant r)$.* Edg_0: *leading term (normal).* Edg_1: *Edgeworth with $n^{-1/2}$ correction term.* Edg_2: *Edgeworth with n^{-1} correction term.*

	$n=4$					$n=8$			
r	*Exact*	Edg_0	Edg_1	Edg_2	r	*Exact*	Edg_0	Edg_1	Edg_2
0	0·0183	0·0401	0·0253	0·0221	2	0·0138	0·0259	0·0160	0·0148
2	0·2381	0·2286	0·2376	0·2394	4	0·0996	0·1079	0·1021	0·1011
4	0·6288	0·5987	0·6289	0·6270	6	0·3134	0·2981	0·3128	0·3141
6	0·8893	0·8944	0·8858	0·8937	8	0·5926	0·5702	0·5926	0·5919
8	0·9786	0·9878	0·9771	0·9780	10	0·8159	0·8116	0·8151	0·8146
10	0·9972	0·9974	0·9478	0·9968	12	0·9362	0·9442	0·9340	0·9374
					14	0·9827	0·9892	0·9820	0·9824

Example 4.3 Sum of independent binary random variables

A more realistic example concerns the sum of independent binary random variables $Y_1, \ldots, Y_n$ with probabilities of 'success' $\theta_1, \ldots, \theta_n$, so that S_n is the total number of 'successes' in n trials. While the generating function for the probability distribution of S_n is easily written down, it is quite often convenient to have a reasonably simple approximation for the cumulative distribution function of S_n and this is provided by the Edgeworth expansion. If we denote the possible values of the binary random variables by 0 and 1, the cumulants of Y_j, are easily shown to be

$$\theta_j, \theta_j(1-\theta_j), \theta_j(1-\theta_j)(1-2\theta_j), \theta_j(1-\theta_j)(1-6\theta_j+6\theta_j^2). \tag{4.19}$$

From these results the cumulants for S_n and hence the corresponding values of ρ_3 and ρ_4 can be found.

For numerical illustration suppose that there are 10 trials with $\theta = \frac{1}{4}$ and two trials with $\theta = \frac{1}{2}$. Table 4.3 compares the exact cumulative distribution function of the number of 'successes' with the leading term of the Edgeworth series, i.e. the normal approximation, and the versions with one and with two correction terms; in all cases a continuity correction was used.

The inclusion of the $n^{-1/2}$ term produces a substantial improvement, but the additional gain from the n^{-1} term is much less apparent.

In section 3.3 we discussed 'amalgamation into the leading term' as a device for choosing that form of an asymptotic expansion which

Table 4.3 *Number X of successes in 12 trials with nonconstant probability of success. Exact distribution function, $P(X \leqslant r)$.* Edg_0, *normal approximation;* Edg_1, *expansion with $n^{-1/2}$ term;* Edg_2, *expansion with n^{-1} term.*

r	*Exact*	Edg_0	Edg_1	Edg_2
0	0·0141	0·0258	0·0187	0·0170
1	0·0892	0·0972	0·0922	0·0950
2	0·2675	0·2582	0·2662	0·2730
3	0·5178	0·5000	0·5170	0·5170
4	0·7498	0·7418	0·7498	0·7430
5	0·9000	0·9028	0·8978	0·8950
6	0·9697	0·9742	0·9671	0·9688
7	0·9932	0·9953	0·9921	0·9935
8	0·9989	0·9994	0·9986	0·9989
9	0·9999	0·9999	0·9998	0·9998

makes the leading term as effective as possible. We can use this idea exactly or approximately in connexion with probability densities, usually by choosing as the leading term a convenient 'familiar' distribution with simple properties.

Example 4.4 Sums of exponential random variables of unequal means

Let $\{X_1, X_2, \ldots\}$ be independent and identically distributed in an exponential distribution of unit mean, let $\{a_1, a_2, \ldots\}$ be a sequence of positive constants and let $Y_j = a_j X_j$, so that Y_j is exponentially distributed with mean a_j. Write $S_n = Y_1 + \cdots + Y_n$ and

$$m_{rn} = (a_1^r + \cdots + a_n^r)/n$$

for the rth moment of $\{a_1, \ldots, a_n\}$.

Then the standardized form for S_n is

$$S_n^* = (S_n - nm_{1n})/\sqrt{(nm_{2m})}. \tag{4.20}$$

Under weak conditions $S_n^* \xrightarrow{\sim} \mathrm{N}(0, 1)$.

Now the support of S_n is the positive real line. Further, if all the a_j are equal then S_n has a gamma distribution. This suggests that, particularly when the a_j vary relatively little, we take a gamma distribution for S_n as the leading term rather than the normal distribution.

To study this in more detail, note that

$$K(Y_j; t) = -\log(1 - a_j t)$$

so that

$$\begin{aligned} K(S_n^*; t) &= -\sqrt{n} m_{1n} t/\sqrt{m_{2n}} - \sum \log\{1 - a_j t/\sqrt{(nm_{2n})}\} \\ &= \tfrac{1}{2}t^2 + \tfrac{1}{3}t^3 \frac{m_{3n}}{m_{2n}^{3/2}\sqrt{n}} + o(n^{-1/2}) \end{aligned} \tag{4.21}$$

if the m_{rn} are bounded as $n \to \infty$. This leads, as above, to a normal limit with a correction term involving the Hermite polynomial of degree 3.

Now suppose that we aim to amalgamate the first two terms of (4.21) into the expansion for a gamma distribution, i.e. to approximate S_n by $c_n X_{d_n}^2$, where $X_{d_n}^2$ has a chi-squared distribution with d_n degrees of freedom. The cumulant generating function of

$$(c_n X_{d_n}^2 - nm_{1n})/\sqrt{(nm_{2n})}$$

is

$$K_\Gamma(t) = -\sqrt{nm_{1n}}t/\sqrt{m_{2n}} - \tfrac{1}{2}d_n \log\left\{1 - \frac{2tc_n}{\sqrt{(nm_{2n})}}\right\}. \qquad (4.22)$$

In choosing c_n and d_n to achieve agreement between (4.21) and (4.22) it is natural, although not essential, to follow the conventional procedure of equating exactly the first two moments, thus ensuring that

$$K_\Gamma(t) = \tfrac{1}{2}t^2 + O(t^3).$$

For this

$$c_n = \tfrac{1}{2}m_{2n}/m_{1n}, \qquad d_n = 2nm_{1n}^2/m_{2n},$$

leading to

$$K_\Gamma(t) = \tfrac{1}{2}t^2 + \tfrac{1}{3}t^3\frac{\sqrt{m_{2n}}}{m_{1n}\sqrt{n}} + O(n^{-1}). \qquad (4.23)$$

To interpret the difference between the expansion (4.21) for $K(S_n^*; t)$ and the expansion for $K_\Gamma(t)$, write, for a particular n,

$$a_j = m_{1n}(1 + a'_{jn}), \qquad \sum a'_{jn} = 0.$$

Then the coefficients of $\frac{1}{3}t^3/\sqrt{n}$ in the two expansions are respectively

$$\frac{1 + 3\sum a'^2_{jn}/n + \sum a'^3_{jn}/n}{(1 + \sum a'^3_{jn}/n)^{3/2}}, \qquad \{\sum(1 + \sum a'^2_{jn}/n)\}^{1/2}. \qquad (4.24)$$

Unless the $\{a'_{jn}\}$ have a substantially negatively skew distribution, the former exceeds the latter: if we expand to the second degree the two expressions are

$$1 + \tfrac{3}{2}\sum a'^2_{jn}/n, \qquad 1 + \tfrac{1}{2}\sum a'^2_{jn}/n, \qquad (4.25)$$

showing that S_n tends to be more skew than the gamma approximation.

Thus amalgamation into the leading term by this route is not totally successful, although (4.25) shows that usually the gamma distribution will be much better as a leading term than the normal approximation which replaces (4.24) and (4.25) by zero.

Of course the argument could be extended to force agreement in the next term, for example by working with an adjustable power of $X^2_{d_n}$.

In most asymptotic calculations the notional quantity becoming large is a sample size or amount of information, but there are other possibilities. As an example we study the effect of a simple form of correlation on the distribution of the standard chi-squared random variable.

Example 4.5 Intra-class correlation and the chi-squared distribution

Suppose that $W_k = Y_1^2 + \cdots + Y_k^2$, where $Y_1, \ldots, Y_k$ are marginally normally distributed with zero mean and unit variance, but are not independent, rather having correlation of the intra-class form, in fact that they have a multivariate normal distribution with $\operatorname{corr}(Y_i, Y_j) = \rho \ (i \neq j)$. It is then convenient for $\rho \geqslant 0$ to write $Y_i = U \sin\theta + Z_i \cos\theta$, where $\{U, Z_1, \ldots, Z_k\}$ are independent standard normal variables and $\rho = \sin^2\theta$. We examine the effect of ρ on the distribution of W_k, supposing ρ to be small.

Now

$$W_k = (k\sin^2\theta + \cos^2\theta)Z^2 + \cos^2\theta \sum (Z_i - \bar{Z})^2,$$

where $\sum (Z_i - \bar{Z})^2$ has a chi-squared distribution with $k-1$ degrees of freedom and independently $Z = (U\sin\theta + \bar{Z}\cos\theta)/\sqrt{(\sin^2\theta + \cos^2\theta/k)}$ has a standard normal distribution. Thus

$$M(W_k; t) = \{1 - 2t(k\sin^2\theta + \cos^2\theta)\}^{-1/2}(1 - 2t\cos^2\theta)^{-(k-1)/2}.$$

If we expand the cumulant generating function in powers of $\rho = \sin^2\theta$, we have that as $\rho \to 0$

$$K(W_k; t) = -\tfrac{1}{2}k\log(1 - 2t) + k(k-1)t^2(1-2t)^{-2}\rho^4 + O(\rho^6),$$

so that

$$\begin{aligned} M(W_k; t) &= (1-2t)^{-k/2}\{1 + k(k-1)t^2(1-2t)^{-2}\rho^4 + O(\rho^6)\} \\ &= (1-2t)^{-k/2}\left[1 + k(k-1)\rho^4\left\{\frac{1}{4} - \frac{1}{2(1-2t)} \right.\right. \\ &\qquad \left.\left. + \frac{1}{4(1-2t)^2}\right\} + O(\rho^6)\right]. \end{aligned}$$

Thus if $q_k(x)$ is the density of the chi-squared distribution with k degrees of freedom, we have on inversion that

$$f(W_k; k) = q_k(x) + \tfrac{1}{4}k(k-1)\rho^4\{q_k(x) - 2q_{k+2}(x) + q_{k+4}(x)\} + O(\rho^6) \tag{4.26}$$

with an exactly corresponding expression for the cumulative distribution function.

The formal connexion with an Edgeworth expansion can be seen by noting that

$$\tfrac{1}{4}k(k-1)\rho^4\{q_k(x) - 2q_{k+2}(x) + q_{k+4}(x)\}$$
$$= \tfrac{1}{4}(k-1)(k+2)^{-1}\rho^4\{k(k+2) - 2x(k+2) + x^2\}q_k(x).$$

The most interesting qualitative feature of this discussion is that the first correction term involves ρ^4 and not, as might have been expected, ρ^2.

A quite different approach that throws some light on the form of the Edgeworth expansion uses a functional equation. Suppose that $\{X_1, X_2, \ldots\}$ is a sequence of independent and identically distributed random variables, copies of a random variable X centred on zero. Let $S_n = X_1 + \cdots + X_n$ and suppose that the standardized form is $S_n^* = S_n/n^a$, where $a = \frac{1}{2}$ for distributions of finite variance. Now for each $r = 1, 2, \ldots$, the sum S_{rn} can be written as the sum of r terms, independent and identically distributed copies of S_n.

To achieve the required generality we work with characteristic functions rather than with moment generating functions, for example with $M(X; it) = E(e^{itX})$. Then, because of the special structure of S_{rn}, we have for each r

$$M(S_{rn}^*; it) = \{M(S_n^*; it/r^a)\}^r. \tag{4.27}$$

Suppose now that as $n \to \infty$, $M(S_n^*; it)$ has an asymptotic expansion

$$M_0(it) + M_1(it)/n^b + o(n^{-b}) \tag{4.28}$$

for some suitable b. Of course it is possible that some function of n other than an inverse power is needed. Then

$$M_0(it) + M_1(it)/(r^b n^b) = \{M_0(it/r^a) + M_1(it/r^a)n^{-b}\}^r + \cdots$$

requiring that

$$M_0(it) = \{M_0(it/r^a)\}^r, \qquad M_1(it) = r^{1+b}\{M_0(it/r^a)\}^{r-1}M_1(it/r^a). \tag{4.29}$$

In (4.29), the constants b and a play quite different roles: a is a characteristic of the limiting law and is determined by the tail behaviour of the density of X, whereas b is concerned with the departure of that tail behaviour from the limiting stable form. Thus for the normal case, $a=\frac{1}{2}$, distributions within the domain of attraction of the normal law fall into a number of classes:

1. those with finite nonzero third cumulant, which have $b=\frac{1}{2}$, $(1+b)/a=3$ and an expansion in which the first correction term is of order $1/\sqrt{n}$ and proportional to the third derivative of $\phi(x)$;
2. those with zero third cumulant and finite nonzero fourth cumulant, which have $b=1$, $(1+b)/a=4$ and an expansion in which the first correction term is of order $1/n$ and proportional to the fourth derivative of $\phi(x)$;
3. obvious extensions of 1 and 2 depending on the first nonzero cumulant;
4. cases in which the third cumulant is infinite (Höglund, 1970).

The first equation in (4.29) is a well known characterization of the stable laws (Renyi, 1970, section 6.8). If $a=\frac{1}{2}$, $M_0(it)=e^{-(1/2)t^2\sigma^2}$, whereas for $a>\frac{1}{2}$ we obtain the characteristic function of the stable law of index $1/a$. In particular, for $a=1$ we have the Cauchy distribution $M_0(it)=e^{-\sigma|t|}$. The second equation has solution

$$M_1(it)=kM_0(it)(it)^{(1+b)/a} \tag{4.30}$$

inverting into a contribution to the density proportional to

$$D_x^{(1+b)/a}\phi^{(a)}(x),$$

where $\phi^{(a)}(x)$ is the density of a stable law of index $1/a$ and $D_x^{(1+b)/a}$ denotes differentiation with respect to x of possibly fractional order $(1+b)/a$; see the end of section 6.11.

Asymptotic expansions associated with other stable laws are considered, for instance, by Cramér (1963). In the special case of the standard Cauchy distribution $(a=1)$ the second term in the expansion will be proportional to $D_x^{1+b}(1+x^2)^{-1}$ and even for integer b this is not of the form of a polynomial multiplying $(1+x^2)^{-1}$.

Rather similar expansions can arise by a quite different route. If we start with a Cauchy distribution centred on zero, with density $\pi^{-1}\lambda(1+\lambda^2x^2)^{-1}$ we can form a compound distribution by supposing λ to be a random variable leading to the density

$$\pi^{-1}E_\Lambda\{\Lambda(1+\Lambda^2x^2)^{-1}\}. \tag{4.31}$$

We can without loss of generality take $E(\Lambda)=1$ and if $\text{var}(\Lambda)=\sigma_\lambda^2$

is small we have the expansion

$$\pi^{-1}(1+x^2)^{-1}+\tfrac{1}{2}\pi^{-1}\sigma_\lambda^2\left[\frac{\partial^2}{\partial\lambda^2}\{\lambda(1+\lambda^2x^2)^{-1}\}\right]_{\lambda=1}+o(\sigma_\lambda^2)$$

$$=\pi^{-1}(1+x^2)^{-1}\left\{1-\sigma_\lambda^2\frac{x^2(3-x^2)}{(1+x^2)^2}+o(\sigma_\lambda^2)\right\}. \tag{4.32}$$

Example 4.6 Generalized inverse Gaussian distribution

For any $\lambda\in R$, $\chi>0$ and $\psi\geqslant 0$, a probability density function on the positive half-line is defined by

$$\frac{(\psi/\chi)^{\lambda/2}}{2K_\lambda\{\sqrt{(\chi\psi)}\}}x^{\lambda-1}e^{-1/2(\chi x^{-1}+\psi x)}, \tag{4.33}$$

K_λ being the modified Bessel function of the third kind with index λ. For $\psi=0$ the norming constant is interpreted as the limiting value for $\psi\downarrow 0$. The distribution given by (4.33) is called the generalized inverse Gaussian distribution. This type of distribution is discussed extensively in Jørgensen (1982). For an account of the most important properties of Bessel functions, including those used below, see Erdélyi, Magnus, Oberhettinger and Tricomi (1953).

Here we will consider only the case where $-1<\lambda<0$, $\chi=1$ and $\psi=0$. In this case, letting $\nu=-\lambda$, (4.33) takes the form

$$2^{-\nu}\Gamma(\nu)^{-1}x^{-\nu-1}e^{-(1/2)x^{-1}} \tag{4.34}$$

and the corresponding moment generating function is

$$2^{1-\nu/2}\Gamma(\nu)^{-1}|t|^{\nu/2}K_\nu\{\sqrt{(2|t|)}\}\quad (t\leqslant 0). \tag{4.35}$$

For $0<\nu<1$, $K_\nu(x)$ possesses the power series expansion

$$K_\nu(x)=\tfrac{1}{2}\Gamma(\nu)\Gamma(1-\nu)\sum_{m=0}^{\infty}\frac{(x/2)^{2m}}{m!}$$
$$\times\left\{\frac{(x/2)^{2-\nu}}{\Gamma(m+3-\nu)}+\frac{(1-\nu)(x/2)^{-\nu}}{\Gamma(m+2-\nu)}-\frac{(x/2)^{\nu}}{\Gamma(m+\nu+1)}\right\} \tag{4.36}$$

as may be shown, for instance, by combining formulae (7.2.12–13) and (7.11.23) of Erdélyi *et al.* (1953).

When $\nu=\tfrac{1}{2}$, (4.34) is the stable law of index $\tfrac{1}{2}$ so from now on we further assume $\nu\neq\tfrac{1}{2}$.

Now, let $X_1,\ldots,X_n$ be a sample from (4.34) and let

$$S_n^* = S_n/n^{1/\nu}.$$

It follows from (4.35) and (4.36) that the cumulant generating function of S_n^* has the asymptotic expansion

$$K(S_n^*;t) = -\frac{\Gamma(1-\nu)}{2^\nu\Gamma(1+\nu)}|t|^\nu - \frac{1}{2}\left\{\frac{\Gamma(1-\nu)}{2^\nu\Gamma(1+\nu)}\right\}^2 |t|^{2\nu}n^{-1} + \frac{1}{2(1-\nu)}|t|n^{1-1/\nu} + \cdots.$$

The form of the leading term shows that the asymptotic distribution of S_n^* is the positive stable law of index ν. (As a check of the calculations, note that the two correction terms cancel if $\nu = \frac{1}{2}$.) It is necessary now to distinguish between the two cases $0 < \nu < \frac{1}{2}$ and $\frac{1}{2} < \nu < 1$. In the notation of formula (4.30) we have:

	k	b	$(1+b)/a$
$0<\nu<\frac{1}{2}$	$-\frac{1}{2}\left\{\frac{\Gamma(1-\nu)}{2^\nu\Gamma(1+\nu)}\right\}^2$	1	2ν
$\frac{1}{2}<\nu<1$	$\frac{1}{2(1-\nu)}$	$1/\nu-1$	1

4.3 Tilted Edgeworth expansion

Equation (4.6), the Edgeworth expansion for a density, shows that the error of the leading term, the standard normal density, is $O(n^{-1/2})$ in general, provided that $\rho_3 \neq 0$. This fact suggests that convergence to normality is relatively slow, especially in the tails of the distribution where the values of $H_3(x)$ will be appreciable.

If, however, we are concerned with behaviour at or near the mean, $x = 0$, the coefficient of $1/\sqrt{n}$ vanishes because $H_3(0) = 0$ and the error of the standard normal approximation is $O(n^{-1})$ rather than $O(n^{-1/2})$. This is well illustrated by the numerical comparisons in Table 4.1. We now consider the exploitation of this by the following device: we connect the density of S_n or S_n^* under the postulated distribution at a particular argument with the density under a modified distribution so chosen that the given argument corresponds to the mean under the new distribution. The Edgeworth expansion is then applied at a point where the error is $O(n^{-1})$ rather than

$O(n^{-1/2})$. A consequence of this procedure is that the resulting approximation has *relative* error $O(n^{-1})$ over a wide range of values of S_n, as contrasted with the direct Edgeworth expansion which has an *absolute*, i.e. additive, error of the indicated order of magnitude.

More specifically, as in section 4.2, let $Y_1, \ldots, Y_n$ be independent and identically distributed random variables, copies of a random variable Y with density $f(y)$, mean μ, variance σ^2, cumulants κ_r and moment and cumulant generating functions $M(t)$ and $K(t)$. We associate with $f(y)$ the exponential family

$$\begin{aligned} f(y; \lambda) &= e^{\lambda y} f(y)/M(\lambda) \\ &= \exp\{\lambda y - K(\lambda)\} f(y), \end{aligned} \tag{4.37}$$

say. The operation of forming $f(y; \lambda)$ is called *exponential tilting*. Note particularly that the moment generating function $M(\lambda)$ for $f(y)$ is exactly the divisor required to normalize the expression $e^{\lambda y} f(y)$.

The key to the subsequent argument lies in the identity

$$f(S_n; s; \lambda) = \exp\{\lambda s - nK(\lambda)\} f(S_n; s; 0), \tag{4.38}$$

where the first argument denotes the random variable and the second the value at which the density is evaluated. Note that this is trivially true for $n = 1$. Further, at each sample point $y^{(n)} = (y_1, \ldots, y_n)$ for which $y_1 + \cdots + y_n = s$, we have because of the independence of the components of $Y^{(n)}$,

$$f(Y^{(n)}; y^{(n)}; \lambda) = \exp\{\lambda s - nK(\lambda)\} f(Y^{(n)}; y^{(n)}; 0). \tag{4.39}$$

Because (4.38) is of key importance we sketch an alternative derivation via moment generating functions (Laplace transforms). In fact

$$E\{\exp(tS_n); \lambda\} = \{E(e^{tY}; \lambda)\}^n = \{M(t + \lambda)/M(\lambda)\}^n,$$

in view of the easily derived form of the moment generating function of Y in (4.37), namely $M(t + \lambda)/M(\lambda)$.

Now if for any function $g(s)$

$$g^\dagger(t) = \int_0^\infty e^{ts} g(s) ds,$$

then

$$g^\dagger(t + \lambda) = \int_0^\infty e^{ts} \{e^{\lambda s} g(s)\} ds,$$

so that $g^{\dagger}(t+\lambda)$ 'inverts' to $e^{\lambda s}g(s)$. Thus $\{M(t)\}^n$ 'inverts' to $f(S_n; s; 0)$, so that $\{M(t+\lambda)\}^n$ 'inverts' to $e^{\lambda s}f(S_n; s; 0)$ and $\{M(t+\lambda)/M(\lambda)\}^n$ to

$$\exp\{\lambda s - nK(\lambda)\}f(S_n; s; 0),$$

which is the required form.

Note that (4.38) can be written in the slightly more general form

$$f(S_n; s; \lambda) = \exp\{(\lambda - \lambda_0)s - nK(\lambda) + nK(\lambda_0)\}f(S_n; s; \lambda_0). \tag{4.40}$$

Suppose now that we wish to calculate a good approximation to $f(S_n; s) = f(S_n; s; 0)$ for a given value of s; typically but not necessarily s will correspond to a fixed value of the standardized random variable S_n^*. We write

$$f(S_n; s) = \exp\{-\lambda s + nK(\lambda)\}f(S_n; s; \lambda).$$

and then choose λ so that s is at the 'centre' of the distribution for that λ. This is achieved by taking $\lambda = \hat{\lambda}$, where

$$E(S_n; \hat{\lambda}) = s.$$

Now $E(S_n; \lambda) = nE(Y; \lambda) = nK'(\lambda)$, so that $\hat{\lambda}$ is defined by

$$K'(\hat{\lambda}) = s/n; \tag{4.41}$$

see section 6.4 for discussion of the existence of $\hat{\lambda}$. Note that $\hat{\lambda}$ has a statistical interpretation as the maximum likelihood estimate of λ from data s under the exponential family model (4.37) generated by $f(y)$ and y. With this choice of λ in (4.38), we have to approximate $f(S_n; s; \hat{\lambda})$ and this we do via the Edgeworth series evaluated at the mean to give

$$f(S_n; s; \hat{\lambda}) = \{2\pi nK''(\hat{\lambda})\}^{-1/2}\{1 + O(n^{-1})\},$$

the leading term arising from the fact that

$$\text{var}(S_n; \hat{\lambda}) = n\,\text{var}(Y; \hat{\lambda}) = nK''(\hat{\lambda}).$$

Thus

$$f(S_n; s; 0) = \frac{\exp\{-\hat{\lambda}s + nK(\hat{\lambda})\}}{\{2\pi nK''(\hat{\lambda})\}^{1/2}}\{1 + O(n^{-1})\}. \tag{4.42}$$

An important feature of (4.42) is that even though its main application will be in a so-called *normal deviation region*

$$|s - nE(Y)| \leqslant c\sqrt{n} \tag{4.43}$$

for fixed c, the formula holds generally for *large deviation regions* of the form $|s - nE(Y)| \leqslant bn$, for fixed b and in certain cases even for all s (see Daniels (1954), Jensen (1988b) and the discussion of exactness cases below). Note that the relative error in the formula is $O(n^{-1})$ as contrasted with the absolute error in (4.6) and moreover that the leading term of (4.42) is never negative.

The formula is thus radically different from the direct Edgeworth series which may behave badly in one or both tails; this point is relevant, for example, if the approximation is to be integrated to calculate a tail area.

The n^{-1} term at the origin in the Edgeworth expansion is given via

$$(3\hat{\rho}_4 - 5\hat{\rho}_3^2)/(24n), \tag{4.44}$$

where for insertion in (4.42) the values of ρ_3 and ρ_4 are calculated at $\hat{\lambda}$.

The leading term of (4.42), considered as function of s, is in general not exactly normalized, i.e. does not integrate to one over the support of S_n. This raises the possibility that we modify (4.42) to

$$f(S_n; s; 0) = c_n \frac{\exp\{-\hat{\lambda}s + nK(\hat{\lambda})\}}{\{2\pi n K''(\hat{\lambda})\}^{1/2}} \{1 + O(n^{-1})\}, \tag{4.45}$$

where c_n is chosen to normalize the leading term. Now if the correction (4.44) is constant, i.e. does not depend on $\hat{\lambda}$, hence on s, then that term will be absorbed into the normalizing constant and the renormalized form will have error that is $O(n^{-3/2})$ uniformly in s. In the great majority of cases, however, (4.44) will vary with s and the resulting normalization will in effect depend on (4.44) in the central region of the distribution and will produce an error that is $O(n^{-3/2})$ only in the region (4.43). There may in some cases be degradation of performance in the tails or, even worse, in more extreme cases the tails may contribute appreciably to c_n thus spoiling performance throughout. These perhaps normally remote possibilities can be studied and to some extent avoided via (4.40). They arise because the region of validity of the expansion is that specified below (4.43). A necessary and sufficient condition that normalization produces a uniform improvement is that (4.44) is constant for all members of the exponential family associated with $f(y)$.

The formula (4.42) is often called the saddlepoint approximation and (4.45) the normalized saddlepoint approximation, the terminology arising from an alternative derivation via the contour integral that inverts the moment generating function of S_n (Daniels, 1954).

Another name is indirect Edgeworth expansion. To emphasize the derivation via exponential tilting, we shall call the expansions *tilted*.

The advantages of the indirect over the direct versions of the Edgeworth expansion are apparent from the above discussion. The main disadvantage lies in the need to evaluate $\hat{\lambda}$ and $K(\lambda)$, although sometimes reasonably convenient approximations can be found to $\hat{\lambda}$ and K.

If the cumulative distribution function is required, term-by-term integration of the direct Edgeworth expansion is immediate. For the tilted expansion a more subtle argument is needed, leading to (4.71) and (4.72) below.

We illustrate the results by some relatively simple examples.

Example 4.7 Normal distribution

If $Y_1, \ldots, Y_n$ are independently normally distributed with mean μ and variance σ^2, then $K(\lambda) = \lambda\mu + \frac{1}{2}\lambda^2\sigma^2$, the cumulant generating function of the normal distribution; the equation (4.41) defining $\hat{\lambda}$ gives

$$\mu + \hat{\lambda}\sigma^2 = s/n$$

and $K''(\hat{\lambda}) = \sigma^2$. Thus the argument of the exponential in (4.42) is, on reduction,

$$-(s - n\mu)^2/(2n\sigma^2)$$

and the leading term of the tilted approximation is exact. The direct Edgeworth expansion also is exact, although only when the correct standardization is used.

Example 4.8 Gamma distribution

We continue the discussion of Example 4.1 in which $Y_1, \ldots, Y_n$ have an exponential distribution of unit mean, so that S_n has the density $s^{n-1}e^{-s}/(n-1)!$ Now $f(y) = e^{-y}$, $M(\lambda) = (1-\lambda)^{-1}$, $K(\lambda) = -\log(1-\lambda)$. The equation (4.41) defining $\hat{\lambda}$ is thus

$$(1-\hat{\lambda})^{-1} = s/n, \qquad -\hat{\lambda}s = n - s,$$

and

$$K(\hat{\lambda}) = \log(s/n), \qquad K''(\hat{\lambda}) = s^2/n^2.$$

The exponential family associated with e^{-y} consists here of exponential distribution of parameter $1-\lambda$.

Thus the tilted approximation (4.42) has leading term

$$\exp\{n - s + n\log(s/n)\}(2\pi n s^2/n^2)^{-1/2} = c_n s^{n-1}e^{-s}, \qquad (4.46)$$

where

$$c_n = (2\pi)^{-1/2}e^n n^{-n+1/2}.$$

This agrees with the exact answer

$$s^{n-1}e^{-s}/\Gamma(n)$$

except for the replacement of $\Gamma(n)$ by the leading term of Stirling's theorem. Thus in this case the normalized form (4.45) is exact. Note that the proportional accuracy of (4.46) is the same for all s, depending only on n. Note also that for the gamma family the quantity (4.44) takes a constant value, essentially because the exponential family is a scale family.

Example 4.9 Exponential family

Let $Y_1,\ldots, Y_n$ be independent and identically distributed random variables, copies of a random variable Y with the exponential family distribution (4.37) with $\lambda = \lambda_0$. Then via (4.40), we have that

$$f(S_n; s; \lambda_0) = \frac{\exp[\{-(\hat{\lambda}-\lambda_0)s + n\{K(\hat{\lambda}) - K(\lambda_0)\}]}{\{2\pi n K''(\hat{\lambda})\}^{1/2}}\{1 + O(n^{-1})\}, \qquad (4.47)$$

with a corresponding normalized form available. Of course this is really just a rewriting of (4.40) with an alternative special form for the distribution of Y.

The normal, gamma and inverse normal distributions are the only ones for which the leading term of the normalized form is exact for all λ_0 in this one-dimensional case.

Example 4.10 von Mises distribution

Suppose that $Z_1,\ldots,Z_n$ are independent and identically distributed random variables, copies of a random variable Z having density

$$\exp(\lambda_0 \cos z)/\{2\pi I_0(\lambda_0)\} \qquad (0 \leqslant z < 2\pi). \qquad (4.48)$$

Thus if we want the distribution of $\sum \cos Z_j$, we first convert (4.48) into a density for $Y = \cos Z$, namely

$$e^{\lambda y}/\{\pi I_0(\lambda_0)(1-y^2)^{1/2}\},$$

so that the exponential family results apply with $K(\lambda) = -\log I_0(\lambda)$. Thus (4.47) gives for the density of $S_n = \sum \cos Z_j$ the leading term

$$\frac{\{I_0(\lambda_0)\}^n}{\{I_0(\hat{\lambda})\}^{n-1}} \frac{e^{s(\hat{\lambda}-\lambda_0)}}{(2\pi n)^{1/2}[\{I_0'(\hat{\lambda})\}^2 - I_0(\hat{\lambda})I_0''(\hat{\lambda})\}]^{1/2}}, \tag{4.49}$$

where $s/n + I_0'(\hat{\lambda})/I_0(\hat{\lambda}) = 0$. These equations can be rewritten using the relations

$$I_0'(\hat{\lambda}) = I_1(\hat{\lambda}), \qquad I_0''(\hat{\lambda}) = I_0(\hat{\lambda}) - I_1(\hat{\lambda})/\hat{\lambda}.$$

From the probabilistic point of view (4.45), or one of its equivalents, is a natural mode of expression. In some statistical applications, however, the exponential family is not merely a convenient mathematical device but is of direct statistical interest. It may then be required to express (4.45) or better (4.47), as a density for $\hat{\Lambda}$, the random variable representing the maximum likelihood estimator of λ. For this we have only to multiply the density of S_n by $ds/d\hat{\lambda} = nK''(\hat{\lambda})$. Therefore the leading term is a modification of the previous one and

$$f(\hat{\Lambda}; \hat{\lambda}; \lambda_0) = \frac{\exp[-(\hat{\lambda}-\lambda_0)s + n\{K(\hat{\lambda}) - K(\lambda_0)\}]}{\sqrt{(2\pi)}} \times \{nK''(\hat{\lambda})\}^{1/2}\{1 + O(n^{-1})\}. \tag{4.50}$$

A more direct statistical interpretation can be obtained as follows. If $L(\lambda; s)$ denotes the likelihood function for λ corresponding to data s, the first part of (4.50) is $L(\lambda_0; s)/L(\hat{\lambda}; s)$, whereas the 'observed information' at the maximum likelihood point is

$$-\left[\frac{\partial^2 \log L(\lambda; s)}{\partial \lambda^2}\right]_{\lambda=\hat{\lambda}} = nK''(\hat{\lambda})$$

and it is convenient to denote this by $\hat{j}_{\lambda\lambda}$. Then (4.50), possibly in normalized form, can be written

$$c\frac{L(\lambda; s)}{L(\hat{\lambda}; s)} \hat{j}_{\lambda\lambda}^{1/2}\{1 + O(n^{-1})\}. \tag{4.51}$$

An important feature of (4.51) is its invariance under reparameterization. That is, if $\phi = \phi(\lambda)$ is a monotone function of λ, so that $\hat{\phi} = \phi(\hat{\lambda})$, we have that

$$f(\hat{\Phi}; \hat{\phi}; \phi) = f_{\hat{\Lambda}}(\hat{\lambda}; \lambda) \Big/ \left| \frac{d\hat{\phi}}{d\hat{\lambda}} \right|. \tag{4.52}$$

But the likelihood function is invariant under reparameterization and

$$\hat{j}_{\phi\phi} = -\left[\frac{\partial^2 \log L}{\partial \phi^2} \right]_{\phi = \hat{\phi}} = \hat{j}_{\lambda\lambda} \left(\frac{d\hat{\lambda}}{d\hat{\phi}} \right)^2. \tag{4.53}$$

Hence (4.51)–(4.53) lead to

$$f(\hat{\Phi}; \hat{\phi}; \phi) = c \frac{L(\phi; s)}{L(\hat{\phi}; s)} \hat{j}_{\phi\phi}^{1/2} \{1 + O(n^{-1})\}, \tag{4.54}$$

showing precise preservation of the form of the leading term.

Example 4.11 Maximum likelihood estimation for the exponential distribution

As an example consider maximum likelihood estimation of the parameter of an exponential distribution written in the forms

$$\lambda e^{-\lambda y}, \ \phi^{-1} e^{-y/\phi}.$$

From observations $y_1, \ldots, y_n$ with mean $\bar{y}$ on independent and identically distributed random variables with this density, we have a likelihood

$$\lambda^n e^{-n\lambda\bar{y}} = \phi^{-n} e^{-n\bar{y}/\phi}$$

and a simple calculation shows that $\hat{\phi} = 1/\hat{\lambda} = \bar{y}$ and that

$$\hat{j}_{\lambda\lambda} = n\bar{y}^2, \qquad \hat{j}_{\phi\phi} = n/\bar{y}^2.$$

Substitution into (4.51) and (4.54) gives for the leading terms

$$c\sqrt{n} e^n \frac{\lambda^n}{\hat{\lambda}^{n-1}} e^{-n\lambda/\hat{\lambda}}, \qquad c\sqrt{n} e^n \hat{\phi}^{n-1} \phi^{-n} e^{-n\hat{\phi}/\phi},$$

which are, except for the normalizing constant, exactly the inverse gamma and gamma densities of $\hat{\lambda} = 1/\bar{y}$ and $\hat{\phi} = \bar{y}$.

Example 4.12 Stable law of index $\frac{1}{2}$

Because of the strong connexion between the derivation of the leading term (4.42) and the Edgeworth series, which in turn is associated with a limiting normal distribution, it is perhaps reasonable to expect (4.42) to give good results only when limiting normality holds. The wider applicability of (4.42) is shown by the stable law of index $\frac{1}{2}$, when (4.42) gives an exact result.

For this let the underlying random variable Y have the density

$$(2\pi)^{-1/2} y^{-3/2} \exp(-\tfrac{1}{2} y^{-1}). \tag{4.55}$$

Then because the random variable S_n/n^2 has a distribution not depending on n, it follows that the density of S_n is

$$n(2\pi)^{-1/2} s^{-3/2} \exp(-\tfrac{1}{2} n^2 s^{-1}). \tag{4.56}$$

To apply (4.42) we write the exponential family associated with (4.55) in the form

$$(2\pi)^{-1/2} y^{-3/2} \exp\{-\tfrac{1}{2} y^{-1} + \lambda y + \surd(-2\lambda)\}. \tag{4.57}$$

That is, $K(\lambda) = -\surd(-2\lambda)$ and it is easily shown that $\hat{\lambda} = -\frac{1}{2} n^2 s^{-2}$, $K''(\hat{\lambda}) = n(-2\hat{\lambda})^{-3/2}$, from which it follows that the leading term (4.42) agrees exactly with (4.56).

Example 4.13 Pure birth process

We now illustrate the application of the tilted approximation to a stochastic process, although in fact the discussion is strongly connected with that for sums of independent random variables. Consider a Markov pure birth process: given j individuals alive at time s, the probability of transition to $j+1$ individuals in $(s, s+\delta s)$ is $\alpha_j \delta s + o(\delta s)$, regardless of the previous history of the process. In the simple birth process $\alpha_j = j\alpha$. Suppose that there is initially one individual.

Two closely related random variables associated with the process are S_n, the time at which the process leaves 'state' n for 'state' $n+1$, i.e. at which the nth new birth occurs; and N_s, the number of individuals at time s. It is clear that

$$f(S_n; s) = \alpha_n P(N_s = n). \tag{4.58}$$

Now S_n is the sum of n independent exponentially distributed random variables with parameters $\alpha_1, \ldots, \alpha_n$ and hence cumulant generating function

$$-\sum_{j=1}^{n} \log(1 + t/\alpha_j). \tag{4.59}$$

The discussion of Example 4.4 could now be applied with $\alpha_j = a_j^{-1}$; this gives an Edgeworth expansion and a connexion with a gamma distribution that is likely to be accurate if the α_j are not too different. Here, however, the α_j are likely to vary strongly.

To apply the present method we introduce the exponential family tilting $\alpha_j \exp(-\alpha_j x)$ to

$$(\alpha_j - \lambda)\exp(\lambda x - \alpha_j x) = \exp\{(\lambda - \alpha_j)x - \log(\alpha_j - \lambda)\}.$$

Thus in the general treatment

$$nK(\lambda) = -\sum_{j=1}^{n} \log(\alpha_j - \lambda), \tag{4.60}$$

the maximum likelihood estimate satisfies

$$s = \sum(\alpha_j - \hat{\lambda})^{-1} \tag{4.61}$$

and

$$nK''(\hat{\lambda}) = \sum(\alpha_j - \hat{\lambda})^{-2}. \tag{4.62}$$

Further, the third and fourth standardized cumulants at $\hat{\lambda}$ are

$$\begin{aligned} \hat{\rho}_3 &= 2\sum(\alpha_j - \hat{\lambda})^{-3}/\{nK''(\hat{\lambda})\}^{3/2}, \\ \hat{\rho}_4 &= 6\sum(\alpha_j - \hat{\lambda})^{-4}/\{nK''(\hat{\lambda})\}^{2}. \end{aligned} \tag{4.63}$$

The leading term of the tilted approximation is thus

$$c_n \frac{e^{-s\hat{\lambda}}}{\prod(\alpha_j - \hat{\lambda})\{2\pi\sum(\alpha_j - \hat{\lambda})^{-2}\}^{1/2}}, \tag{4.64}$$

where c_n is a normalization constant. Introduction of the next term multiplies this by

$$\left\{1 + \frac{3\hat{\rho}_4 - 5\hat{\rho}_3^2}{24n}\right\}. \tag{4.65}$$

By (4.58), this is easily converted into an approximation for the distribution of N_s.

Daniels (1982), to whom this argument is due, gave numerical results for the simple epidemic, that is the special case in which $\alpha_j = j(N - j + 1)$, where N is the initial number of susceptibles and there is initially one infected individual. Table 4.4 gives a brief extract from his results showing the quite high relative accuracy of the leading term, the improvement achieved by including the correction (4.65) and the maintenance of a small relative error in the tails of

Table 4.4 *Simple epidemic, $N = 5$. Probability distribution of number N_s at time s. (a) Exact; (b) renormalized saddlepoint, leading term; (c) renormalized saddlepoint, second approximation.*

n	$s=$	0·1	0·4	0·8	1·6
1	(a)	0·6065	0·1354	0·01832	0·0003355
	(b)	0·6174	0·1428	0·01992	0·0003623
	(c)	0·6053	0·1345	0·01819	0·0003403
2	(a)	0·2620	0·1576	0·02776	0·0005555
	(b)	0·2561	0·1583	0·02861	0·0005786
	(c)	0·2628	0·1576	0·02748	0·0005584
3	(a)	0·09641	0·1809	0·04636	0·001087
	(b)	0·09299	0·1795	0·04681	0·001105
	(c)	0·09675	0·1812	0·04620	0·001090
4	(a)	0·02841	0·1986	0·08938	0·002995
	(b)	0·02722	0·1967	0·09012	0·002965
	(c)	0·02851	0·1988	0·08959	0·003015
5	(a)	0·005973	0·1940	0·2116	0·01956
	(b)	0·005700	0·1917	0·2146	0·01963
	(c)	0·005995	0·1940	0·2114	0·01991
6	(a)	0·0006741	0·1336	0·6066	0·9756
	(b)	0·0006415	0·1310	0·5999	0·9754
	(c)	0·0006767	0·1338	0·6071	0·9751

the distribution. He also discusses the consequence of normalization, namely the achievement of overall improvement at some cost at particular points.

Integration over some region of the tilted approximation (4.42) or its normalized form (4.45) will, under mild regularity conditions, lead to approximations of the probability of that region which are accurate to orders $O(n^{-1})$ and $O(n^{-3/2})$, respectively. For some discussions of the type of regularity conditions involved, see Daniels (1954), Durbin (1980) and Jensen (1988b). Only rarely can such integration be carried out explicitly, and one then has to resort to numerical integration or to approximate or asymptotic methods, for instance Laplace's method.

As a case of particular statistical interest, suppose an approximation is sought to $P(S_n \leqslant s_0)$ based on the normalized tilted approximation (4.45), i.e. we wish to evaluate

$$\int_{-\infty}^{s_0} \{c_n/\sqrt{(2\pi)}\}\{nK''(\hat{\lambda})\}^{-1/2} e^{-\hat{\lambda}s + nK(\hat{\lambda})} ds, \tag{4.66}$$

where $\hat{\lambda}$ is determined as a function of s by

$$nK'(\hat{\lambda}) = s. \tag{4.67}$$

Note that, by (4.67), (4.66) can be transformed to

$$\int_{-\infty}^{\lambda_0} \{c_n/\sqrt{(2\pi)}\}\{nK''(\hat{\lambda})\}^{1/2} e^{-n\{\hat{\lambda}K'(\hat{\lambda})-K(\hat{\lambda})\}} d\hat{\lambda}, \tag{4.68}$$

where $K'(\lambda_0) = s_0$. (We are assuming here, for simplicity, that both s and $\hat{\lambda}$ range over the entire real line. In case the range of one or both of them is an interval obvious modifications are necessary.) If (4.68) cannot be calculated explicitly, which is usually the case, we may proceed by making a further transformation by introducing the variable

$$q = \operatorname{sgn}(\hat{\lambda})[2\{\hat{\lambda}K'(\hat{\lambda}) - K(\hat{\lambda})\}]^{1/2}. \tag{4.69}$$

Since the derivative of $\lambda K'(\lambda) - K(\lambda)$ is $\lambda K''(\lambda)$ the transformation from $\hat{\lambda}$ to q is monotonically increasing, and (4.68) transforms to

$$\int_{-\infty}^{q_0} \{c_n\sqrt{n}/\sqrt{(2\pi)}\}(q/\hat{\lambda})\{K''(\hat{\lambda})\}^{-1/2} e^{-(n/2)q^2} dq. \tag{4.70}$$

This integral is of the type (3.75) with $Q_\tau(\infty) = 1$ and hence, by (3.78) and noting that $c_n = 1 + O(n^{-1})$, we have on letting $r = \sqrt{n}q$ and $v = \hat{\lambda}/\{nK''(\hat{\lambda})\}^{1/2}$ and dropping the suffixes 0 and n,

$$P(S \leqslant s) \doteq \Phi(r') + \left(\frac{1}{r'} - \frac{1}{v'}\right)\phi(r'), \tag{4.71}$$

with error $O(n^{-1})$ uniformly in s.

Note that r and v have the interpretation, respectively, of being the signed likelihood ratio statistic and the score statistic for testing $\lambda = 0$ in the exponential tilt model determined by S_n.

The formula (4.71) is for continuous variables S_n. In the lattice case a slight modification is needed, the resulting expression being

$$P(S \leqslant s) \doteq \Phi(r') + \left(\frac{1}{r'} - \frac{1}{v'}\right)\phi(r'), \tag{4.72}$$

where

$$v' = \{K''(\hat{\lambda}')\}^{1/2}\{\exp(\tfrac{1}{2}\hat{\lambda}') - \exp(-\tfrac{1}{2}\hat{\lambda}')\} \tag{4.73}$$

and $\hat{\lambda}'$ and r' are determined by

$$nK'(\hat{\lambda}') = s - \tfrac{1}{2} \tag{4.74}$$

and $r' = r(\hat{\lambda}')$, i.e. we work with a 'continuity corrected saddlepoint'. For a detailed derivation see Daniels (1987).

Example 4.14 Terminating renewal process

We now outline a rather different application of exponential tilting. For a more detailed discussion, see Feller (1966, Ch. 11). Let $g(x)$ be the density of a nonnegative improper random variable X, i.e.

$$\gamma = \int_0^\infty g(x)dx < 1, \tag{4.75}$$

so that formally there is a probability $1-\gamma$ that $X = \infty$. Now provided that

$$\int_0^\infty e^{\lambda x} g(x)dx$$

converges for sufficiently large λ, there exists a unique $\tilde{\lambda} > 0$ such that

$$\int_0^\infty e^{\tilde{\lambda} x} g(x)dx = 1, \tag{4.76}$$

so that $\tilde{g}(x) = e^{\tilde{\lambda} x} g(x)$ is the density of a proper random variable, $\tilde{X}$, say. This is a very special case of exponential tilting.

Let $\{X_1, X_2, \ldots\}$ and $\{\tilde{X}_1, \tilde{X}_2, \ldots\}$ be sequences of independent and identically distributed random variables, copies respectively of X and of $\tilde{X}$. They define renewal processes, i.e. sequences of points at, in the first case, $X_1, X_1 + X_2, \ldots$ The first process is terminating in the sense that after N steps and at time T, where

$$P(N = n) = (1-\gamma)\gamma^n \qquad (n = 0, 1, \ldots),$$

an infinite X_{n+1} is encountered and the process disappears to infinity. Denote the renewal densities of the two processes by $h(t)$ and $\tilde{h}(t)$, so that, for example, $h(t)\delta t + o(\delta t)$ is the probability of a point in the first process falling in the interval $(t, t+\delta t)$. It is a familiar result in renewal theory that these densities satisfy integral equations, namely

$$\begin{aligned} h(t) &= g(t) + \int_0^t h(t-u)g(u)du, \\ \tilde{h}(t) &= \tilde{g}(t) + \int_0^t \tilde{h}(t-u)\tilde{g}(u)du. \end{aligned} \tag{4.77}$$

It follows immediately that $\tilde{h}(t) = e^{\tilde{\lambda}t}h(t)$. Further, because the second process is a proper renewal process, it is known that $\tilde{h}(t) \to 1/\tilde{\mu}$ as $t \to \infty$, where $\tilde{\mu} = E(\tilde{X})$. Thus

$$h(t) \sim \tilde{\mu}^{-1}e^{-\tilde{\lambda}t}. \tag{4.78}$$

Now

$$\int_a^b h(u)du$$

is the expected number of points in (a, b), so that it follows that M_t, the number of 'proper' points in (t, ∞), is such that

$$E(M_t) \sim (\tilde{\mu}\tilde{\lambda})^{-1}e^{-\tilde{\lambda}t}. \tag{4.79}$$

But

$$E(M_t) = P(T > t)E(M_t | T > t) = P(T > t)/(1 - \gamma),$$

i.e. for large t

$$P(T > t) \sim (1 - \gamma)(\tilde{\mu}\tilde{\lambda})^{-1}e^{-\tilde{\lambda}t}. \tag{4.80}$$

It is easily shown that in the special case $g(x) = \phi e^{-x}$, for $0 < \phi < 1$, $\gamma = \phi = 1 - \tilde{\lambda}$, and the last result is exact.

4.4 Cornish–Fisher inversion

We now return to the direct Edgeworth expansion of section 4.2, in particular to the formula (4.7) for the distribution function of the standardized form, S_n^*. Let $k_{n\alpha}$ be the upper α quantile, i.e. the solution for fixed α of the equation

$$F(S_n^*; k_{n\alpha}) = \alpha. \tag{4.81}$$

Now as $n \to \infty$, $k_{n\alpha} \to k_\alpha$, where $\Phi(k_\alpha) = \alpha$ and $\Phi(z)$ is the standardized normal distribution function. The most direct approach is thus to solve

$$\Phi(k_{n\alpha}) - \phi(k_{n\alpha})\{\rho_3 H_2(k_{n\alpha})/(6\sqrt{n}) + \rho_4 H_3(k_{n\alpha})/(24n) + \rho_3^2 H_5(k_{n\alpha})/(72n)\} + O(n^{-3/2}) = \alpha \tag{4.82}$$

by the technique of section 3.5. This yields

$$k_{n\alpha} = k_\alpha + \frac{1}{6\sqrt{n}}(k_\alpha^2 - 1)\rho_3 + \frac{1}{24n}(k_\alpha^3 - 3k_\alpha)\rho_4 - \frac{1}{36n}(2k_\alpha^3 - 5k_\alpha)\rho_3^2 + O(n^{-3/2}). \tag{4.83}$$

This is called the Cornish–Fisher inversion of the Edgeworth expansion.

An equivalent approach is via a stochastic expansion of the random variable S_n. By the argument of section 3.8 it follows that in the light of (4.82) we can write, for example,

$$S_n^* = Z + \frac{1}{6\sqrt{n}}(Z^2 - 1)\rho_3 + O_p(n^{-1}), \tag{4.84}$$

where Z has a standardized normal distribution. Then, in the region in which (4.82) is a monotone function of Z, quantiles of S_n^* can be found by substituting into (4.84) the corresponding quantile of Z. The monotonicity requirement is needed also in (4.83) and corresponds to the condition that the density function is nonnegative: it will, of course, be satisfied for sufficiently large n in any case.

Example 4.15 Gamma distribution

It is useful to use yet again the gamma distribution in illustration. In order to make direct comparisons with widely available tables, we consider the chi-squared distribution with n degrees of freedom via the random variable

$$W_n = X_1^2 + \cdots + X_n^2,$$
$$W_n^* = (W_n - n)/\sqrt{(2n)},$$

where $X_1, \ldots, X_n$ are independent and identically distributed copies of a random variable X with a standard normal distribution and W_n^* is the standardized form of W_n. We have that $\rho_3(X^2) = 2\sqrt{2}$, $\rho_4(X^2) = 12$. Equation (4.83) now gives approximations to the α quantile of W_n^* corresponding to those for the corresponding quantiles of W_n. The leading term and two correction terms are

$$n + \sqrt{(2n)}\left\{k_\alpha + \frac{\sqrt{2}}{3\sqrt{n}}(k_\alpha^2 - 1) + \frac{1}{18n}(k_\alpha^3 - 7k_\alpha)\right\}, \tag{4.85}$$

where k_α is the α quantile for the standardized normal distribution.

Table 4.5 makes some comparisons with the exact quantiles of the chi-squared distribution. Convergence is remarkably quick, especially so since the originating random variables X^2 being added are far from normal. The first correction term makes a substantial improvement on the normal approximation. It is interesting, and of

Table 4.5 *Accuracy of Cornish–Fisher expansion for chi-squared distribution with n degrees of freedom.* Edg_0, *normal approximation;* Edg_1, *with* $n^{-1/2}$ *correction term;* Edg_2, *with* n^{-1} *correction term.*

	$\alpha = 0{\cdot}01$				$\alpha = 0{\cdot}1$			
n	*Exact*	Edg_0	Edg_1	Edg_2	*Exact*	Edg_0	Edg_1	Edg_2
5	15·09	12·36	15·20	15·07	9·24	9·65	9·48	9·24
10	23·21	20·40	23·34	23·25	15·99	15·73	16·16	15·99
50	76·15	73·26	76·20	76·16	63·17	62·82	63·24	63·16
100	135·81	132·90	135·84	135·81	118·50	118·12	118·55	118·50

	$\alpha = 0{\cdot}9$				$\alpha = 0{\cdot}99$			
n	*Exact*	Edg_0	Edg_1	Edg_2	*Exact*	Edg_0	Edg_1	Edg_2
5	1·61	0·95	1·38	1·62	0·55	−2·35	0·59	0·72
10	4·87	4·27	4·70	4·87	2·56	−0·08	2·86	2·95
50	37·67	37·18	37·61	37·69	29·71	26·74	29·68	29·72
100	82·36	81·88	82·30	82·35	70·06	67·10	70·04	70·07

some general significance, that while the second correction term usually makes a further improvement, it worsens the approximation in the region where the approximation is rather bad, that is in the lower tail when the degrees of freedom are small.

Example 4.16 Sensitivity of quantiles to nonnormality

Some simple statistical procedures use averages and differences of particular sample quantiles (order statistics), e.g. the interquantile range. It is possible to regard these as of intrinsic interest, i.e. as defining aspects of the underlying distribution regardless of any assumption such as normality. Alternatively we may link such quantities with the mean and standard deviation of the underlying distribution. This is best done via the expected values of the order statistics but a rather easier approximate argument uses the corresponding distributional quantiles as approximations via (4.83).

For this we have from (4.83) that the α and $1-\alpha$ quantiles of the distribution of S_n are given in terms of the mean, standard deviation, skewness and kurtosis, μ', σ', ρ_3', ρ_4', of S_n by

$$\mu' + \sigma'\left\{k_\alpha + \frac{\rho_3'}{6}(k_\alpha^2 - 1) + \frac{\rho_4'}{24}(k_\alpha^3 - 3k_\alpha) - \frac{\rho_3'^2}{36}(2k_\alpha^3 - 5k_\alpha)\right\},$$

$$\mu' + \sigma'\left\{-k_\alpha + \frac{\rho_3'}{6}(k_\alpha^2 - 1) - \frac{\rho_4'}{24}(k_\alpha^3 - 3k_\alpha) + \frac{\rho_3'^2}{36}(2k_\alpha^3 - 5k_\alpha)\right\}. \quad (4.86)$$

Now for $\rho_3' \neq 0$ the average of these is, to the indicated approximation, equal to μ' if and only if $k_\alpha^2 = 1$, i.e. $\alpha = 0{\cdot}84$. The condition that the difference is $2k_\alpha\sigma'$ depends on $\rho_3'^2/\rho_4'$ but if attention is restricted to symmetrical distributions the condition is $k_\alpha^2 = 3$, i.e. $\alpha = 0{\cdot}96$.

Inversions analogous to the Fisher–Cornish expansion can be carried out when the leading term in the expansion for the distribution function is nonnormal and in especially convenient form when the quantiles of that distribution have simple form.

Example 4.17 Distribution of minima

A convenient example for which exact and asymptotic solutions can be compared is provided by the distribution of the minimum of uniform random variables. Let $Y_1, \ldots, Y_n$ be independently uniformly distributed on (0, 1) and let $M_n = \min(Y_1, \ldots, Y_n)$. Then

$$\begin{aligned} F(M_n; x) &= P(M_n \leqslant x) = 1 - P(M_n > x) \\ &= 1 - (1 - x)^n. \end{aligned}$$

The α quantile, $m_{n\alpha}$, is thus defined by $F(M_n; m_{n\alpha}) = \alpha$, i.e.

$$m_{n\alpha} = 1 - (1 - \alpha)^{1/n}. \quad (4.87)$$

Now in the corresponding asymptotic theory we write $M_n^* = nM_n$ when

$$\begin{aligned} F(M_n^*; z) &= P(M_n \leqslant z/n) \\ &= 1 - (1 - z/n)^n \\ &= 1 - e^{-z}\left(1 - \frac{z^2}{2n}\right) + O(n^{-2}). \end{aligned} \quad (4.88)$$

Thus if we want the α quantile of M_n^* we solve

$$e^{-z_{\alpha n}}\left\{-\frac{z_{\alpha n}^2}{2n} + O(n^{-2})\right\} = 1 - \alpha$$

starting with $z_\alpha = -\log(1-\alpha)$. It follows that

$$z_{n\alpha} = z_\alpha - \frac{z_\alpha^2}{2n} + O(n^{-2}), \tag{4.89}$$

so that $m_{n\alpha} = z_{n\alpha}/n$ has the expansion

$$z_\alpha/n - z_\alpha^2/(2n^2) + O(n^{-3}), \tag{4.90}$$

as it is easily verified directly from (4.87).

In conclusion, note that an approximation to the cumulative distribution function of S_n or S_n^* can be produced in various ways, directly from the Edgeworth expansion, or via the Cornish–Fisher inversion giving, for example,

$$P(S_n^* \leqslant s) \doteq \Phi(z),$$

where

$$s = z + (z^2 - 1)\rho_3/(6\sqrt{n}),$$

or via amalgamation into the leading term which to this order amounts to replacing z by

$$s - (s^2 - 1)\rho_3/(6\sqrt{n}).$$

Beard, Pentikäinen and Pesonen (1984, section 3.11) discuss an application to risk theory where the second of these gives much more accurate answers than the first.

4.5 Nonlinear functions of sums

In the previous sections we have discussed in some detail asymptotic expansions associated with sums of independent and identically distributed random variables. The discussion can be generalized in various ways, for example to sums of nonidentically distributed random variables. Under suitable restrictions on the third and fourth cumulants, as well as the usual requirements on the variances, a direct generalization of the results on Edgeworth expansions is available; see p. 93.

In the present section, however, we discuss a different generalization useful, in particular, in statistical applications, applying where a nonlinear function of a sum of independent random variables is involved; for example, the maximum likelihood estimate is in simple problems such a nonlinear function.

Suppose then that T_n is a random variable converging in probability to θ and that $\sqrt{n}(T_n - \theta)$ is not only asymptotically normal but has an Edgeworth expansion. Let $g(t)$ be a function of t well-behaved at $t = \theta$. We want an asymptotic expansion for $g(T_n)$. Of course, a direct approach is to take the expansion for T_n and to divide by $|g'(t_n)|$, expressing the answer in terms of the new variable. The discussion of maximum likelihood estimates and tilted approximations, especially in (4.52)–(4.54) illustrates this.

One approach to more explicit results is via the stochastic expansion of random variables outlined in section 3.8. By this, because of the assumed Edgeworth expansion we can write

$$\sqrt{n}(T_n - \theta) = X + a_1(X)/\sqrt{n} + a_2(X)/n + O_p(n^{-3/2}),$$

where the a's are polynomials and X is a random variable with a normal distribution of zero mean. It now follows on direct substitution that

$$g(T_n) = g\{\theta + X/\sqrt{n} + a_1(X)/n + a_2(X)/n^{3/2} + O_p(n^{-2})\}.$$

Taylor expansion then gives

$$\begin{aligned}\sqrt{n}\{g(T_n) - g(\theta)\} = {} & Xg'(\theta) + n^{-1/2}\{a_1(X)g'(\theta) + \tfrac{1}{2}X^2 g''(\theta)\} \\ & + n^{-1}\{a_2(X)g'(\theta) + Xa_1(X)g''(\theta) \\ & + \tfrac{1}{6}X^3 g'''(\theta)\} + O_p(n^{-3/2}).\end{aligned}$$

This has exactly the form of the stochastic expansion of a random variable whose distribution has an Edgeworth expansion with coefficients determined from the standard expansions for the first four moments of T_n.

In other words, nonlinear functions of random variables with Edgeworth expansions themselves have such expansions. In particular, this applies to maximum likelihood estimates.

Example 4.18 Nonlinear functions of normal random variables of small variance

Suppose that T_n is exactly normally distributed with variance σ^2/n and mean $\theta > 0$. Then in the above discussion $a_1 = a_2 = 0$ and

$$\begin{aligned}\sqrt{n}\{g(T_n) - g(\theta)\} &= Xg'(\theta) + \tfrac{1}{2}n^{-1/2}X^2 g''(\theta) + \tfrac{1}{6}n^{-1}X^3 g'''(\theta) + \cdots \\ &= X\{g'(\theta) - \tfrac{1}{2}n^{-1/2}g''(\theta) + \tfrac{1}{2}n^{-1}g'''(\theta)\} \\ &\quad + \tfrac{1}{2}n^{-1/2}g''(\theta)H_2(X) + \tfrac{1}{6}n^{-1}g'''(\theta)H_3(X) + \cdots\end{aligned}$$

showing that the Edgeworth expansion applies to nonlinear functions of exactly (or approximately) normal random variables as well as to linear functions of nonnormal random variables.

An interesting special case is $g(x) = x^2$, when we recover by yet another route some properties of the noncentral chi-squared distribution with one degree of freedom.

Further results and exercises

4.1 The random variable X_n has a distribution tending to the chi-squared distribution with d degrees of freedom as $n \to \infty$. Denote the density of that distribution by $q_d(x)$. Show that the following expansions are formally equivalent.

(a) The density of X_n is

$$q_d(x)(1 - a/n) + q_{d+2}(x)an^{-1} + O(n^{-2}).$$

(b) The density of X_n is

$$q_d(x)\{1 + a(x/d - 1)n^{-1}\} + O(n^{-2}).$$

(c) The moment generating function of X_n is

$$(1 - 2t)^{-(1/2)d}\{1 + 2at(1 - 2t)^{-1}n^{-1}\} + O(n^{-2}).$$

(d) The density of $\{1 + 2a/(dn)\}X_n$ is $q_d(x)$, with error $O(n^{-2})$. Suppose next that (a) is replaced by

$$q_d(x)(1 - a/n - b/n) + aq_{d+2}(x)/n + bq_{d+4}(x)/n + O(n^{-2}).$$

Examine the analogues of (b) and (c). Show that if $b \neq 0$ amalgamation into a chi-squared density cannot be achieved by a scaling constant and an adjustment to the degrees of freedom.

[Section 4.2]

4.2 A stationary linear time series $\{Y_n\}$ of zero mean is defined by

$$Y_n = \sum_{m=-\infty}^{\infty} b_m U_{n-m},$$

where $\{U_s\}$ are independent and identically distributed random variables of zero mean. Suppose that the coefficients $\{b_m\}$ satisfy

$$\sum_{m=-\infty}^{\infty} |mb_m| < \infty, \qquad \sum_{m=-\infty}^{\infty} b_m \neq 0;$$

in particular this implies that the process has short range dependence. The question of Edgeworth expansions for

$$Y_1 + \cdots + Y_n$$

can now be addressed by rewriting the sum in terms of the U_s. If U_s has a finite cumulant of order k and some power of its characteristic function is integrable, then an Edgeworth expansion with $k-1$ correction terms is available.

[Section 4.2; Skovgaard, 1986]

4.3 Let $\{Y_1, Y_2, \ldots\}$ be a stationary Gaussian process of zero mean and unit variance and let $g(Y)$ be a function of zero mean and finite variance. To extend the central limit theorem consider

$$Z_n = \frac{1}{c_n} \sum_{i=1}^{n} g(Y_i),$$

where c_n^2 is asymptotically proportional to

$$\operatorname{var}\left\{ \sum_{i=1}^{n} g(Y_i) \right\}.$$

For simplicity we take $c_n = n^b$, where $\frac{1}{2} \leqslant b < 1$. Now if $b = \frac{1}{2}$, the process has short-range dependence and it can be shown that a normal limit is obtained, as in the special case of independence. If, however, $\frac{1}{2} < b < 1$ a nonnormal limit may be achieved, even though the individual random variables have finite variance. A key role in this is played by the Hermite rank of g, that is, the smallest value of r such that $E\{g(Y)H_r(Y))\} \neq 0$. In fact to determine the limiting distributions it is enough to take $g(y) = H_r(y)$. If $r = 1$, a normal limit is achieved, but if $r = 2$, and $b > 3/4$ the limiting form is not normal. For a thorough discussion in terms of the convergence of the process $\{Z_i\}$, see Taqqu (1974).

[Section 4.2]

4.4 In the discussion of Example 4.5 show that the intra-class correlation ρ is such that $\rho \geqslant -1/(k-1)$. Does the discussion in terms of the angular representation for Y_i hold for $\rho < 0$? Develop corresponding expansions when $Y_1, \ldots, Y_k$ are generated by a stationary Gaussian first-order autoregressive process of zero mean and small correlation.

[Section 4.2]

4.5 In some statistical applications it may be desired to approximate a sampling distribution of zero mean and unit variance by an Edgeworth series depending on, say, standardized third and fourth cumulants which have to be estimated from data, e.g. from independent simulated data. Discuss the effect of such errors of estimation on the adequacy of the resulting approximation.

[Section 4.2]

4.6 In Edgeworth expansions for asymptotically normal statistics it will commonly happen that the odd-order cumulants have expansions in terms of powers of $n^{-1/2}$ and the even-order cumulants in powers of n^{-1}. This is true also in the case of multivariate statistics.

[Section 4.2; Withers, 1984]

4.7 In the careful mathematical study of Edgeworth and other expansions an important role is played by the Esséen smoothing lemma. Let F and G be two univariate cumulative distribution functions such that (i) F has zero mean, (ii) $F(x) - G(x)$ tends to zero as $x \to \pm\infty$, (iii) G has a density g bounded by m, (iv) the Fourier transform (characteristic function) of g is continuously differentiable with value 1 and first derivative zero at the origin. Then if $\Delta(\gamma)$ is the difference of the characteristic functions of F and G at argument γ, we have

$$|F(x) - G(x)| \leqslant \pi^{-1} \int_{-T}^{T} |\Delta(\gamma)/\gamma| d\gamma + 24m/(\pi T).$$

[Section 4.2; Feller, 1966, p. 511; Esséen, 1945]

4.8 In such nonergodic stochastic processes as the pure birth process and the explosive Ornstein–Uhlenbeck process the observed information (minus the second derivative of the log likelihood) divided by its expectation tends not to a constant but to a random variable. A natural route to the study of asymptotic expansions for such problems is thus to condition on the observed information: it is then found that expansions closely related to those for independent and identically distributed random variables can be achieved. In general the unconditional expansions involve mixtures of normal distributions and associated functions.

[Section 4.2; Jensen, 1988a]

4.9 Let Y_1, $Y_2, \ldots$ be independent and identically distributed random vectors, copies of an $m \times 1$ random vector $(Y^1, \ldots, Y^m)^T$. Define minima component by component, i.e. let $Z_n^j = \min(Y_1^j, \ldots, Y_n^j)$. Suppose that there exist normalizing constants c_n^j, $d_n^j > 0$ such that $X_n^1 = (Z_n^1 - c_n^1)/d_n^1, \ldots, X_n^m = (Z_n^m - c_n^m)/d_n^m$ has a limiting survivor function $H(x^1, \ldots, x^m) = \lim P(X_n^1 > x_n^1, \ldots, X_n^m > x_n^m)$.

(i) Prove that H has the minimum stabilizing property that for each $r = 1, 2, \ldots$ there exist constants $k_r^1, \ldots, k_r^m$, $l_r^1, \ldots, l_r^m$ such that

$$H^r(x^1, \ldots, x^m) = H(k_r^1 x^1 + l_r^1, \ldots, k_r^m x^m + l_r^m). \tag{4.91}$$

(ii) Show that the marginal survivor function of each component has the minimum stabilizing property and hence has one of the classical extreme value forms.

(iii) Show that a function H with marginal unit exponential survivor functions is produced by the following construction. Let $Z_1, \ldots, Z_m$ be independent and identically distributed unit exponential random variables and let $a_{ij} \geqslant 0, \sum_i a_{ij} = 1$ $(j = 1, \ldots, m)$. Write $X^j = \min(Z_1/a_{1j}, \ldots, Z_m/a_{mj})$. Then

$$\begin{aligned} H(x^1, \ldots, x^m) &= \exp\left\{-\left(\sum_i \max_j (a_{ij} x^j)\right)\right\} \\ &= \exp\left\{-\int_{\mathscr{S}_m} \max_j (u^j x^j) d\mu(u)\right\}, \end{aligned} \tag{4.92}$$

where $\mu(\cdot)$ is a finite measure on the simplex $\mathscr{S}_m$:

$$\mathscr{S}_m = \{(u^1, \ldots, u^m) : u^j \geqslant 0, \sum u^j = 1\}.$$

By taking limits show that the required property holds for (4.92) with $\mu(\cdot)$ any positive measure. (In fact all minimum stabilizing distributions with unit exponential margins are of the form (4.92).)

(iv) Examine asymptotic expansions associated with (4.91).

(v) Verify that for $n = 2$, with components (x, y), the survivor functions

(a) $\exp\{-x - y + \theta xy/(x + y)\}$ $\quad (0 \leqslant \theta \leqslant 1)$,
(b) $\exp\{-(x^\theta + y^\theta)^{1/\theta}\}$ $\quad (1 \leqslant \theta)$,

are of the above form.

(vi) Show how by changes of sign and transformation, results can be derived for maxima rather than minima and for extreme-value marginals other than the exponential.

[Section 4.2; de Haan and Resnick, 1977; Leadbetter, Lindgren and Rootzén, 1983, Tiago de Oliveira, 1984]

4.10 If the Edgeworth series is continued to a large number of terms, large oscillations may occur in the tails of the distribution; indeed the Edgeworth expansion considered as an infinite series may well diverge. The oscillations are, however, to a considerable degree suppressed by an initial transformation of the statistic to near symmetry.

[Sections 4.2, 4.5; Niki and Konishi, 1986]

4.11 Let $Y_1, \ldots, Y_n$ be independent and identically distributed random variables, copies of a random variable Y having density $\lambda e^{-\lambda y}$ or equivalently $\mu^{-1}e^{-y/\mu}$. Write down from previous results the Edgeworth expansion up to the term in n^{-1} for the distribution function of $\bar{Y} = \sum Y_j/n$, the maximum likelihood estimate of μ. Hence deduce from first principles the existence of and then calculate the Edgeworth expansion for $1/\bar{Y}$, the maximum likelihood estimate of λ.

[Sections 4.2, 4.5]

4.12 Examine the tilted approximation for the sum of independent binomial random variables with the same probability of success.

[Section 4.3]

4.13 In simple permutation tests for the comparison of two groups from matched pairs, we use random variables of the form

$$W_n = \sum_{j=1}^{n} a_j Y_j$$

where $Y_1, Y_2, \ldots$ are independent and identically distributed random variables, copies of a random variable Y with $P(Y = 1) = P(Y = -1) = \frac{1}{2}$, and $\{a_j\}$ are given positive constants with well-behaved values of $S_{nr} = \sum_{j=1}^{n} a_j^r/n$. Prove that the cumulant generating function of W_n is

$$K(W_n; t) = \sum \log\cosh(a_j t)$$

and hence obtain a normal approximation and Edgeworth expansion for the distribution function of W_n involving $\operatorname{var}(W_n) = ns_{n2}$ and the standardized fourth cumulant $-2s_{n4}/(ns_{n2}^2)$.

Obtain a formal tilted approximation to the distribution of W_n via the above cumulant generating function.

Note that in general W_n has a complicated support: approximations to a cumulative distribution function can be justified in the usual way but approximations to a density have to be taken in a rather formal sense.

[Section 4.3; Bartlett, 1935; Robinson, 1982]

4.14 In polymer physics there is some interest in a two- (or three-) dimensional random walk with n steps, the steps being of preassigned lengths $l_1,\ldots,l_n$, the directions of the steps being independently uniformly distributed. Then, in two dimensions, the projections on the x-axis of the distance between the two ends of the walk is

$$Y_n = \sum_{j=1}^{n} l_j \cos \Theta_j,$$

where $\{\Theta_j\}$ are independently uniformly distributed on $(0, 2\pi)$.

Show that the cumulant generating function of Y_n is

$$K(Y_n; t) = \sum_{j=1}^{n} \log I_0(tl_j)$$

and hence obtain the tilted approximation for the density of Y_n.

Discuss why, if special interest attaches to the region $y \doteq \sum l_j$, an Edgeworth expansion is unlikely to be effective. It can be shown also that the tilted approximation can reproduce multimodal behaviour. For further details, numerical results and further references, see Weiss and Kiefer (1983).

[Section 4.3]

4.15 Prove that the only univariate distributions for which the normalized tilted approximation for the sum of n independent and identically distributed values is exact are the normal, the inverse Gaussian and the gamma distributions.

(i) For this first check by direct calculations that the three distributions indeed have the stated property.

(ii) Next show that the two next terms in the tilted approximation involves $\alpha_1 = 3\rho_4 - 5\rho_3^2$ and $\alpha_2 = 24\rho_6 + 105\rho_4^2 + 168\rho_3\rho_5 - 630\rho_3^2\rho_4 + 385\rho_3^4$, where the standardized cumulants are evaluated at the 'maximum likelihood' point.

(iii) Hence note that α_1 and α_2 (and subsequent terms) must be independent of the maximum likelihood point.

(iv) Show that $\alpha_1 = 0$ leads to the normal and inverse Gaussian distribution, exact without normalization. For this obtain a differential equation in terms of the cumulant generating function.

(v) The next possibility, $\alpha_1 = c \neq 0$, $\alpha_2 = c'$, where c, c' are constants, leads after considerable manipulation to the condition that all standardized cumulants are constants and hence to the normal or gamma distribution.

[Section 4.3; Blæsild and Jensen, 1985]

4.16 Show that in a Poisson process of rate ρ, the probability $p(y, t)$ that there are y points in $(0, t)$ satisfies

$$\frac{\partial p(y,t)}{\partial t} = \rho p(y-1,t) - \rho y p(y,t)$$

and that if this is converted into a partial differential equation by regarding the step from $y-1$ to y as small (relative to the range of variation) there results the diffusion equation

$$\frac{\partial p(y,t)}{\partial t} = -\rho\frac{\partial p(y,t)}{\partial y} + \tfrac{1}{2}\rho\frac{\partial^2 p(y,t)}{\partial y^2}$$

with solution the normal approximation to the Poisson distribution. Show that if only the first-order term in $\partial/\partial y$ is included, the 'deterministic' approximation results.

Show further that if the expansion is made instead in terms of $l(y,t) = \log p(y,t)$ the first-order approximation is $\partial l/\partial t = \rho\{\exp(-\partial l/\partial y) - 1\}$ and yields the Poisson distribution with $y!$ replaced by Stirling's formula. See Daniels (1960) who discusses similar approximations for a general class of time-homogeneous processes and the connexions with the tilted method.

[Section 4.4]

4.17 Examine the $n^{-1/2}$ terms in the Edgeworth expansion for s^2, s, $s^{2/3}$, $\log s$, where s^2 is a normal theory estimate of variance of σ^2 with m degrees of freedom, i.e. ms^2/σ^2 has the chi-squared distribution with m degrees of freedom.

[Section 4.5]

4.18 If $Y_1, \ldots, Y_n$ are independent and identically distributed random variables, copies of a random variable Y having probability density $f(y;\theta)$, let the function $h(y,\theta)$ be such that for all θ, $E\{h(Y,\theta)\} = 0$ and for all $y, h(y,\theta)$ is monotone decreasing in θ. An estimate $\tilde{\theta}$ of θ is then provided by the unique root of

$$\sum_{j=1}^{n} h(Y_j, \theta) = 0,$$

and the equation is said to be an unbiased estimating equation. Show that $\tilde{\theta} > a$ if and only if $\sum h(Y_j, a) > 0$ and hence obtain Edgeworth and tilted approximations for the distribution of $\tilde{\theta}$.

For alternative methods and numerical comparisons, see Field and Hampel (1982) and Daniels (1983).

[Section 4.5]

Bibliographic notes

Edgeworth expansions have a long history; see Cramér (1937) and Feller (1966, Ch. 16) for proofs of their asymptotic character. Cornish and Fisher (1937) studied their inversion. Exponential tilting as a device for generating asymptotic expansions is due to Esscher (1932) and the systematic use of the tilted method in a probabilistic context to Daniels (1954); for a critical review and bibliography, see Reid (1988). Example 4.16 is due to Benson (1949).

CHAPTER 5

Miscellany on multivariate distributions

5.1 Introduction

In the previous chapters we have developed some key ideas for univariate random variables. In dealing with extensions to multivariate random variables it is natural to start in two dimensions, partly because that case is of intrinsic interest and partly because the bivariate formulae provide a natural introduction to the formalism needed to handle the general multivariate case concisely.

In two dimensions we define the moment generating function of $X=(X_1,X_2)$ by

$$\begin{aligned} M(X;t) &= M(t) = E(e^{t\cdot X}) \\ &= E(e^{t_1X_1+t_2X_2}) \\ &= 1+(\mu'_{10}t_1+\mu'_{01}t_2) \\ &\quad +\frac{1}{2!}\mu'_{20}t_1^2+\mu'_{11}t_1t_2+\frac{1}{2!}\mu'_{02}t_2^2+\cdots, \end{aligned} \tag{5.1}$$

where $\mu'_{r_1r_2}=E(X_1^{r_1}X_2^{r_2})$ is a moment about the origin. This is a generalization of (1.12) in the univariate case.

A similar definition applies for moments about the mean $\mu=(\mu'_{10},\mu'_{01})$,

$$\begin{aligned} \mu_{r_1r_2} &= E\{(X_1-\mu'_{10})^{r_1}(X_2-\mu'_{01})^{r_2}\}, \\ M(X-\mu;t) &= E(e^{t\cdot(X-\mu)}). \end{aligned} \tag{5.2}$$

To express this in a concise notation we write $r=(r_1,r_2)$, $|r|=r_1+r_2$, $t^r=t_1^{r_1}t_2^{r_2}$, $r!=r_1!r_2!$,

$$\mu'_r=E(X^r)=E(X_1^{r_1}X_2^{r_2}) \tag{5.3}$$

and then put the expansion (5.1) in the form

$$M(t) = \sum_{\nu=0}^{\infty} \sum_{|r|=\nu} \mu'_r \frac{t^r}{r!} \tag{5.4}$$

with a corresponding expression for the moments about the mean. Note that we group together terms of the same degree $\nu = |r|$.

This notation extends immediately to the m-dimensional case, where with $X = (X_1, \ldots, X_m)$, $r = (r_1, \ldots, r_m)$, $t^r = t_1^{r_1} \cdots t_m^{r_m}$, $r! = r_1! \cdots r_m!$, formula (5.4) continues to apply.

We now define the cumulant generating function and cumulants by direct analogy with the one-dimensional case. That is,

$$\begin{aligned} K(X;t) = K(t) &= \log M(t) \\ &= \sum_{\nu=1}^{\infty} \sum_{|r|=\nu} \kappa_r \frac{t^r}{r!}. \end{aligned} \tag{5.5}$$

For the multivariate normal distribution $\kappa_r = 0$, $|r| > 2$.

Exactly as in the one-dimensional case

$$K(X - \mu; t) = -t \cdot \mu + K(X; t),$$

so that the second and higher degree cumulants depend only on the moments about the mean. Now for bivariate distributions and for fairly low-order cumulants it is simple, if mildly tedious, to expand the expressions

$$\begin{aligned} \log(1 &+ \tfrac{1}{2}\mu_{20}t_1^2 + \mu_{11}t_1t_2 + \tfrac{1}{2}\mu_{02}t_2^2 + \tfrac{1}{6}\mu_{30}t_1^3 \\ &+ \tfrac{1}{2}\mu_{21}t_1^2t_2 + \tfrac{1}{2}\mu_{12}t_1t_2^2 + \tfrac{1}{6}\mu_{03}t_2^3 + \cdots) \\ &= \tfrac{1}{2}\kappa_{20}t_1^2 + \kappa_{11}t_1t_2 + \tfrac{1}{2}\kappa_{02}t_2^2 + \tfrac{1}{6}\kappa_{30}t_1^3 \\ &+ \tfrac{1}{2}\kappa_{21}t_1^2t_2 + \tfrac{1}{2}\kappa_{12}t_1t_2^2 + \tfrac{1}{6}\kappa_{03}t_2^3 + \cdots \end{aligned}$$

and the corresponding exponential forms to obtain the relations

$$\begin{aligned} &\kappa_{20} = \mu_{20}, \quad \kappa_{11} = \mu_{11}, \quad \kappa_{02} = \mu_{02}, \\ &\kappa_{30} = \mu_{30}, \quad \kappa_{21} = \mu_{21}, \quad \kappa_{12} = \mu_{12}, \quad \kappa_{03} = \mu_{03}, \\ &\kappa_{40} = \mu_{40} - 3\mu_{20}^2, \quad \kappa_{31} = \mu_{31} - 3\mu_{20}\mu_{11}, \\ &\kappa_{22} = \mu_{22} - \mu_{20}\mu_{02} - 2\mu_{11}^2, \quad \kappa_{13} = \mu_{13} - 3\mu_{02}\mu_{11}, \\ &\kappa_{04} = \mu_{04} - 3\mu_{02}^2 \end{aligned}$$

with the μ's given in terms of the κ's as

$$\mu_{40} = \kappa_{40} + 3\kappa_{20}^2, \quad \mu_{31} = \kappa_{31} + 3\kappa_{20}\kappa_{11},$$
$$\mu_{22} = \kappa_{22} + \kappa_{20}\kappa_{02} + 2\kappa_{11}^2, \quad \mu_{13} = \kappa_{13} + 3\kappa_{02}\kappa_{11},$$
$$\mu_{04} = \kappa_{04} + 3\kappa_{02}^2.$$

The complexity of these formulae increases rapidly with their degree.

Note that if a is an $m \times m$ matrix of constants

$$K(aX;t) = K(X;a^{\mathrm{T}}t);$$

we study linear transformations in more detail in section 5.5.

As in the one-dimensional case it is often convenient to introduce the dimensionless forms

$$\rho_{r_1r_2} = \kappa_{r_1r_2}/(\kappa_{20}^{1/2r_1}\kappa_{02}^{1/2r_2}).$$

It is clear, however, that in more dimensions and for higher-order cumulants and moments some more powerful technique is needed to handle these expansions. In section 5.3 we explain some notation and further definitions that aid this, but first in section 5.2 we discuss briefly some special cases.

5.2 Special properties

For particular distributions the moments may be calculated directly or, when feasible, via the appropriate generating function. For mixed moments it may be simpler to proceed via conditional expectations:

$$E(X_1^{r_1}X_2^{r_2}) = E\{X_1^{r_1}E(X_2^{r_2}|X_1)\}.$$

Example 5.1 Bivariate log normal distribution

Suppose that $X = (X_1, X_2)$ has a bivariate normal distribution with mean (μ_1, μ_2) and variances σ_{20}, σ_{02} and covariance σ_{11}. Then $Y = (Y_1, Y_2) = (e^{X_1}, e^{X_2})$ is said to have a bivariate log normal distribution. From the moment generating function of X we have that

$$\begin{aligned} E(Y_1^{r_1}Y_2^{r_2}) &= E\{\exp(r_1X_1 + r_2X_2)\} \\ &= \exp(r_1\mu_1 + r_2\mu_2 + \tfrac{1}{2}r_1^2\sigma_{20} + r_1r_2\sigma_{11} + \tfrac{1}{2}r_2^2\sigma_{02}). \end{aligned} \quad (5.6)$$

In particular we have the univariate results that

$$E(Y_1) = \exp(\mu_1 + \tfrac{1}{2}\sigma_{20}), \qquad \operatorname{var}(Y_1) = \exp(2\mu_1 + \sigma_{20})(e^{\sigma_{20}} - 1). \quad (5.7)$$

Further, the conditional distribution of X_2 given $X_1 = x_1$ being normal with mean and variance

$$\mu_2 + (\sigma_{11}/\sigma_{20})(x_1 - \mu_1), \qquad \sigma_{02\cdot 1} = \sigma_{02} - \sigma_{11}^2/\sigma_{20},$$

it follows that the conditional distribution of Y_2 given $Y_1 = y_1$ is log normal with mean and variance

$$\begin{aligned} &(y_1/\mu_1)^{\sigma_{11}/\sigma_{20}} \exp(\mu_2 + \tfrac{1}{2}\sigma_{02\cdot 1}), \\ &(y_1/\mu_1)^{2\sigma_{11}/\sigma_{20}} \exp(2\mu_2 + \sigma_{02\cdot 1})\{\exp(\sigma_{02\cdot 1}) - 1\}, \end{aligned} \tag{5.8}$$

thus showing nonlinearity of regression and proportional change in the conditional standard deviation, as is clear from the structure of the random vector.

We can now from (5.6) find moments about the origin, hence moments about the mean, and hence cumulants and standardized cumulants.

The cumulants up to third order are

$$\begin{aligned} \kappa_{11} &= \operatorname{cov}(Y_1, Y_2) = \exp(\mu_1 + \mu_2 + \tfrac{1}{2}\sigma_{20} + \tfrac{1}{2}\sigma_{02})(e^{\sigma_{11}} - 1), \\ \kappa_{30} &= \exp(3\mu_1 + \tfrac{3}{2}\sigma_{20})(e^{\sigma_{20}} - 1)^2(e^{\sigma_{20}} + 2), \\ \kappa_{21} &= \exp(2\mu_1 + \mu_2 + \sigma_{20} + \tfrac{1}{2}\sigma_{02})(e^{\sigma_{11}} - 1)(e^{\sigma_{20}+\sigma_{11}} + e^{\sigma_{20}} - 2), \end{aligned} \tag{5.9}$$

with similar expressions for κ_{12} and κ_{03}. For interpretation they are best standardized to give

$$\begin{aligned} \rho_{11} &= \operatorname{corr}(Y_1, Y_2) = (e^{\sigma_{11}} - 1)/\{(e^{\sigma_{20}} - 1)(e^{\sigma_{02}} - 1)\}^{1/2}, \\ \rho_{30} &= (e^{\sigma_{20}} - 1)^{1/2}(e^{\sigma_{20}} + 2), \\ \rho_{21} &= \frac{(e^{\sigma_{11}} - 1)(e^{\sigma_{20}+\sigma_{11}} + e^{\sigma_{20}} - 2)}{(e^{\sigma_{20}} - 1)(e^{\sigma_{02}} - 1)^{1/2}}. \end{aligned} \tag{5.10}$$

One route to simplification and further interpretation is to suppose $\sigma_{ij} = \sigma'_{ij}/t$ and to let $t \to \infty$.

In some problems it may be easier to work with an underlying structure.

Example 5.2 A latent linear structure

Suppose that U, V, W are independent univariate random variables with cumulant generating functions $K(U; t)$, $K(V; t)$, $K(W; t)$ and let

$$X_1 = U + W, \qquad X_2 = V + W. \tag{5.11}$$

Then

$$M(X;t)=M(U;t_1)M(V;t_2)M(W;t_1+t_2),$$

so that

$$K(X;t)=K(U;t_1)+K(V;t_2)+K(W;t_1+t_2). \qquad (5.12)$$

On expansion it follows that the mixed cumulants are such that

$$\kappa_r(X)=\kappa_{r_1+r_2}(W) \qquad (r_1,r_2\neq 0), \qquad (5.13)$$

because the mixed cumulants arise only from W.

Equation (5.13) applied to low-order cumulants could be the basis for testing conformity with (5.11). Note that the simple result (5.13) applies to cumulants and not to moments.

Important special cases of (5.11) arise when U, V, W have Poisson distributions, or binomial distributions with the same probabilities of 'success', or gamma distributions with the same rate parameter.

In most cases we shall use the cumulants and moments as tools for the calculation of asymptotic expansions for distribution functions. As such, beyond the general interpretation that the dimensionless cumulants $\rho_{r_1r_2}$ are measures of departure from multivariate normality, it may not be necessary to give individual moments and cumulants a direct interpretation.

To a limited extent, however, we can generalize one interpretation of the correlation coefficient ρ_{11}, namely that $1-\rho_{11}^2$ is the proportion of the variance of X_2 unaccounted for by linear least squares regression on X_1, with a symmetrical interpretation if the roles of X_2 and X_1 are interchanged.

For this generalization consider the linear least squares regression of X_2 on X_1 and X_1^2. Without loss of generality take $E(X)=0$ and then write

$$X_2=\beta_1X_1+\beta_2(X_1^2-\mu_{30}X_1/\mu_{20}-\mu_{20})+\varepsilon_{2,1}, \qquad (5.14)$$

where the quadratic term has been orthogonalized with respect to the linear term and arranged to have zero mean and $\varepsilon_{2,1}$ is a residual. A standard least squares calculation gives

$$\beta_1=\mu_{11}/\mu_{20}, \qquad \beta_2=(\mu_{21}-\mu_{30}\mu_{11})/(\mu_{40}-\mu_{30}^2/\mu_{20}-\mu_{20}^2)$$

and

$$\operatorname{var}(\varepsilon_{2,1})=\mu_{02}-\frac{\mu_{11}^2}{\mu_{20}}-\frac{(\mu_{21}-\mu_{30}\mu_{11})^2}{\mu_{40}-\mu_{30}^2/\mu_{20}-\mu_{20}^2}.$$

In terms of standardized cumulants,

$$\operatorname{var}(\varepsilon_{2,1}) = \mu_{02}\left\{1 - \rho_{11}^2 - \frac{(\rho_{21} - \rho_{30}\rho_{11})^2}{2 + \rho_{40} - \rho_{30}^2}\right\}. \qquad (5.15)$$

In particular, if X_1 is symmetrically distributed, $\rho_{30} = 0$, and

$$\operatorname{var}(\varepsilon_{2,1}) = \mu_{02}\{1 - \rho_{11}^2 - \rho_{21}^2/(2 + \rho_{40})\}, \qquad (5.16)$$

giving a fairly direct interpretation of ρ_{21}^2 in the same spirit as that of ρ_{11}^2. The contribution of X_1^2 adjusting for X_1 is zero if and only if $\rho_{21} = 0$. On interchanging the roles of X_1 and X_2, ρ_{12}^2 has a similar interpretation.

We have also from (5.15) that

$$1 - \rho_{11}^2 - (\rho_{21} - \rho_{30}\rho_{11})^2/(2 + \rho_{40} - \rho_{30}^2) \geqslant 0$$

with equality if and only if X_2 is with probability one a quadratic function of X_1, emphasizing that the mixed cumulants, as indeed the marginal cumulants, cannot be assigned arbitrarily.

5.3 Index notation

Judicious use of index notation, adopting and developing classical conventions of coordinate-based differential geometry, is of great help in many statistical calculations when the random variate and/or the parameter are multivariate. In fact, employing such notation, the multivariate versions of the formulae are often no more complicated than the univariate, and univariate calculations sometimes become more transparent by viewing them as special cases of the corresponding multivariate. The gain is very considerable already within the scope of the present monograph, and it turns out to be even more pronounced when it comes to applications in statistical theory.

We use primarily $i, j, k, \ldots$, without or with suffixes, to denote indices, each such index running from 1 to m for some positive integer m. Further, we adopt the Einstein summation convention according to which if an index occurs both as a superscript and as a subscript in a single expression then summation over that index is understood. Thus, for instance, $a^i b_i = a^1 b_1 + \cdots + a^m b_m$ and $a^{ijk} b_{il} = a^{1jk} b_{1l} + \cdots + a^{mjk} b_{ml}$. As another example, of particular interest, the summation convention allows us to write Taylor's formula for a function f of

m variables compactly as

$$f(x+t) = \sum_{\nu=0}^{\infty} \frac{1}{\nu!} t^{i_1} \cdots t^{i_\nu} f_{/i_1 \cdots i_\nu}(x), \tag{5.17}$$

where $x = (x^1, \dots, x^m)$, $t = (t^1, \dots, t^m)$, $f_{/i_1 \cdots i_\nu}$ denotes the νth order derivative of f with respect to $x^{i_1}, \dots, x^{i_\nu}$ and by the summation convention the sum is over all possible choices of $i_1, \dots, i_\nu$. Even more compactly we may write

$$f(x+t) = \sum_{|I|=0}^{\infty} \frac{1}{|I|!} t^I f_{/I}, \tag{5.18}$$

where I stands for an index set $i_1 \cdots i_\nu$ and $|I| = \nu$, while $t^I = t^{i_1} \cdots t^{i_\nu}$.

We speak of quantities with upper and/or lower indices as *multiarrays*, and upper indices are said to be *contravariant* while lower indices are called *covariant*. The distinction is determined by behaviour under transformation, as we shall see later. Occasionally we use $r, s, t, \dots,$ rather than $i, j, k, \dots$ as indices. Further, from now on, we mostly index random variables by superscripts.

As an example, let $X^1, \dots, X^m$ be m random variables and let $\kappa^I = \kappa^{i_1 \cdots i_\nu}$ denote the joint moment of $X^{i_1}, \dots, X^{i_\nu}$ for some set $I = i_1 \cdots i_\nu$ of ν indices, i.e.

$$\kappa^I = E(X^{i_1} \cdots X^{i_\nu}).$$

Then κ^I is a multiarray with ν contravariant indices. Note the change in notation from the bivariate case.

Now, for every fixed value of r and i, let c_i^r be a real number. These numbers constitute a two-dimensional array with one covariant index and one contravariant index, and we may think of this as an $m \times m$ matrix. Defining $Y^r = c_i^r X^i (= c_1^r X^1 + \cdots + c_m^r X^m)$ we have that the vector $(Y^1, \dots, Y^m)$ is a linear transformation of $(X^1, \dots, X^m)$ by the matrix c_i^r and, furthermore, that if we write

$$\kappa^{r_1 \cdots r_\nu} = E(Y^{r_1} \cdots Y^{r_\nu}),$$

then

$$\kappa^{r_1 \cdots r_\nu} = c_{i_1}^{r_1} \cdots c_{i_\nu}^{r_\nu} \kappa^{i_1 \cdots i_\nu}.$$

(It will almost always be clear from the choice of letters which random variables are involved. If this is not the case, however, the notation is easily extended by writing, for example, $\kappa^{r_1 \cdots r_\nu}(Y)$, etc.) Any quantity A, say, which obeys this type of transformation law relative to underlying linear transformations, i.e.

$$A^{r_1\cdots r_\nu} = c^{r_1}_{i_1}\cdots c^{r_\nu}_{i_\nu} A^{i_1\cdots i_\nu}$$

for any nonsingular matrix $c = [c^r_i]$, is termed a *Cartesian tensor*, more specifically a contravariant Cartesian tensor of rank ν, with a similar definition of covariant Cartesian tensors, namely that

$$A_{r_1\cdots r_\nu} = c^{i_1}_{r_1}\cdots c^{i_\nu}_{r_\nu} A_{i_1\cdots i_\nu}$$

for any nonsingular matrix $c = [c^i_r]$. Note that, by contrast to many treatments of Cartesian tensors, the matrices c are not restricted to being orthogonal. The covariance matrix of $X^1,\ldots,X^m$ which we denote by $\kappa^{i,j}$, i.e.

$$\kappa^{i,j} = \kappa^{ij} - \kappa^i\kappa^j = \mathrm{cov}\,(X^i, X^j),$$

is a contravariant Cartesian tensor, and that the precision matrix, i.e. the inverse of $\kappa^{i,j}$, which we denote by $\kappa_{i,j}$, is a covariant Cartesian tensor, in symbols

$$Y^r = c^r_i X^i, \qquad X^i = c^i_r Y^r, \qquad \kappa^{r,s} = c^r_i c^s_j \kappa^{i,j}, \qquad \kappa_{r,s} = c^i_r c^j_s \kappa_{i,j},$$

where c^i_r denotes the matrix inverse to c^r_i. The easiest way to check the result for $\kappa_{r,s}$ is to verify that if $\kappa_{i,j}$ is the inverse of $\kappa^{i,j}$, so that $\kappa_{i,j}\kappa^{j,k} = \delta^k_i$, then $\kappa_{r,s}$ defined as above is indeed the inverse of $\kappa^{r,s}$.

There are two simple useful operations known as *lowering and raising of indices*. To illustrate these, suppose $\phi^{i,j}$ is a matrix with inverse $\phi_{i,j}$ and let, for instance, a^k and b_{lm} be multiarrays. Then we say that the array $a_k = \phi_{k,k'}a^{k'}$ has been obtained by lowering the contravariant index k (using ϕ) and $b^{lm} = \phi^{l,l'}\phi^{m,m'}b_{l'm'}$ is said to be derived by raising of the indices. The essential point is that if $\phi^{i,j}$, a^k, b_{lm} are Cartesian tensors then so are $\phi_{i,j}$, a_k and b^{lm}, each obeying its appropriate transformation law, and similarly for general arrays.

Let $I = i_1\cdots i_\nu$ be an arbitrary index set. We call ν the *size* of I and, as above, denote this by $|I|$. For any $\alpha = 1,\ldots,\nu$ we let $I_* = I_1,\ldots,I_\alpha$ denote a partition of I into α blocks. Suppose that F is a mapping from the set of all partitions of I into the reals, i.e. F defines a real number for each partition. (Partitions are discussed in section 5.8.) Further, for given α,

$$\sum_{I/\alpha} F(I_1,\ldots,I_\alpha) \tag{5.19}$$

will indicate the sum of $F(I_1,\ldots,I_\alpha)$ over all partitions of I into α

blocks. (Note that the order in which we write the blocks carries no significance here.)

It is essential that in determining the possible partitions of I the elements $i_1, \ldots, i_\nu$ are considered as distinct, even though some of them may have the same index value. Thus, for instance, if $I = ijk$ and $\alpha = 2$ the partitions in question are

$$i, jk \quad j, ik \quad k, ij,$$

and these are considered as formally different even in cases like $i = j = k = 1$ or $i = 1, j = k = 3$. In the applications we have in mind the *value* of F will, however, not depend on such formal distinctions.

Occasionally, and in particular when writing out special cases, we shall express (5.19) alternatively in a way which can most easily be explained by example. Thus

$$\sum_{ijk/2} F(I_1, I_2) = F(i, jk)[3],$$

$$\sum_{ijkl/2} F(I_1, I_2) = F(i, jkl)[4] + F(ij, kl)[3],$$

$$\sum_{ijkl/3} F(I_1, I_2, I_3) = F(i, j, kl)[6],$$

etc. In general, $[n]$ after a symbol will indicate a sum of n similar terms, determined by suitable permutations of the indices. Note that since, as mentioned above, the functions F we shall consider are such that their values do not reflect the formal distinction between identical index values, we have relations such as

$$\sum_{111/2} F(I_1, I_2) = 3F(1, 11),$$

$$\sum_{133/2} F(I_1, I_2) = F(1, 33) + 2F(3, 13),$$

etc.

Using these notational conventions we may, in particular, give compact expression to a formula for the higher order derivatives of a composite function $g \circ f$, where f and g are real functions of variables x and y, respectively, with $x = (x^1, \ldots, x^m) \in R^m$, and $y \in R$. In the notation of (5.18) and (5.19)

$$(g \circ f)_{/I}(x) = \sum_{\alpha=1}^{|I|} g^{(\alpha)}(y) \sum_{I/\alpha} f_{/I_1}(x) \cdots f_{/I_\alpha}(x) \tag{5.20}$$

for $y=f(x)$, $g^{(\alpha)}(y)=d^\alpha g/dy^\alpha$ and for any index set I. When $m=1$ this may be rewritten as

$$(g\circ f)^{(v)}(x)=\sum_{\alpha=1}^{v} g^{(\alpha)}(y)\sum\frac{v!}{n_1!\cdots n_k!}\{f^{(r_1)}(x)/r_1!\}^{n_1}\cdots\{f^{(r_k)}(x)/r_k!\}^{n_k}$$

where the inner sum is over all positive integers $k, n_1,\ldots,n_k$ and $r_1,\ldots,r_k$ such that no two r_j are the same and

$$n_1+\cdots+n_k=\alpha, \qquad n_1r_1+\cdots+n_kr_k=v.$$

The latter formula is known as Faà di Bruno's formula; see also Lemma 5.1 in section 5.9.

5.4 The exlog relations

Let $t=(t_1,\ldots,t_m)$ be an m-dimensional vector variable and for $I=i_1\cdots i_\nu$ an arbitrary index set, let

$$t_I=t_{i_1\cdots i_\nu}=t_{i_1}\cdots t_{i_\nu}$$

be the ordinary product of the relevant components of t.

Now, consider a power series

$$p_1(t)=1+\sum_{\nu=1}^{\infty}\frac{1}{\nu!}c^{i_1\cdots i_\nu}t_{i_1}\cdots t_{i_\nu}, \tag{5.21}$$

where we use the summation convention. The power series may be formal in that we are not concerned with its convergence. Let

$$p_0(t)=\sum_{\nu=1}^{\infty}\frac{1}{\nu!}c^{i_1,\ldots,i_\nu}t_{i_1}\cdots t_{i_\nu} \tag{5.22}$$

be the corresponding power series for $p_0(t)=\log p_1(t)$. The coefficients $c^{i_1\cdots i_\nu}$ and $c^{i_1,\ldots,i_\nu}$ are assumed all to be symmetric under arbitrary permutations of the indices. Thus we assume that the coefficients in (5.21) and (5.22) are related as if the two series were ordinary convergent power series (and that $p_1(t)>0$). We wish to describe the relations between the two sets of coefficients c, and we will refer to the formulae (5.27), (5.28), (5.31), (5.32), (5.34) and (5.35) derived below as the *exlog relations*.

For this it is convenient to introduce the notation

$$c_1^{i_1\cdots i_\nu}=c^{i_1\cdots i_\nu}, \tag{5.23}$$

$$c_0^{i_1\cdots i_\nu}=c^{i_1,\ldots,i_\nu}. \tag{5.24}$$

This allows us to rewrite (5.21) and (5.22) as

$$p_1(t) = 1 + \sum_{|I|=1}^{\infty} \frac{1}{|I|!} c_1^I t_I \tag{5.25}$$

and

$$p_0(t) = \sum_{|I|=1}^{\infty} \frac{1}{|I|!} c_0^I t_I, \tag{5.26}$$

where, as usual, I is a generic symbol for an index set $i_1 \cdots i_v$ of size $|I| = v$, v being an arbitrary positive integer.

More importantly, we can express the coefficients of $p_1(t)$ in terms of those of $p_0(t)$ simply as

$$c_1^I = \sum_{\alpha=1}^{|I|} \sum_{I/\alpha} c_0^{I_1} \cdots c_0^{I_\alpha}, \tag{5.27}$$

the inner summation being as defined by (5.19).

Further (5.27) may be solved for c_0^I to yield

$$c_0^I = \sum_{\alpha=1}^{|I|} (-1)^{\alpha-1}(\alpha-1)! \sum_{I/\alpha} c_1^{I_1} \cdots c_1^{I_\alpha}. \tag{5.28}$$

The fact that the coefficients c_1 and c_0 are related by formulae (5.27) and (5.28) follows directly from formula (5.20), on taking f to be exp or log in the latter.

It is, however, simple to check the validity of (5.27) for the first few values of I. Written out the first few instances of (5.27) are

$$\begin{aligned} c^{ij} &= c^{i,j} + c^i c^j, \\ c^{ijk} &= c^{i,j,k} + c^i c^{j,k}[3] + c^i c^j c^k, \\ c^{ijkl} &= c^{i,j,k,l} + c^i c^{j,k,l}[4] + c^{i,j} c^{k,l}[3] + c^i c^j c^{k,l}[6] + c^i c^j c^k c^l, \end{aligned} \tag{5.29}$$

the corresponding inverse relations being

$$\begin{aligned} c^{i,j} &= c^{ij} - c^i c^j, \\ c^{i,j,k} &= c^{ijk} - c^i c^{jk}[3] + 2c^i c^j c^k, \\ c^{i,j,k,l} &= c^{ijkl} - c^i c^{jkl}[4] - c^{ij} c^{kl}[3] + 2c^i c^j c^{kl}[6] - 6c^i c^j c^k c^l, \end{aligned} \tag{5.30}$$

cf. (5.28).

It turns out to be convenient to introduce generalized c-coefficients by

$$c^{I_1,\ldots,I_\alpha} = \sum_{I_* \leqslant I'_*} (-1)^{\beta-1}(\beta-1)! c_1^{I'_1} \cdots c_1^{I'_\beta}. \tag{5.31}$$

Here $I_1,\ldots,I_\alpha$ and $I'_1,\ldots,I'_\beta$ both indicate partitions of a given index set I, and the summation is over all partitions $I'_* = I'_1,\ldots,I'_\beta$ of $I = I_1\cdots I_\alpha$ such that $I_* = I_1,\ldots,I_\alpha$ is a subpartition of I'_*. Note that the definition (5.31) is consistent with the notation (5.23) and (5.24); cf. (5.28).

It will be shown in section 5.9 by the method of Möbius inversion that formula (5.27) generalizes to

$$c^{I_1,\ldots,I_\alpha} = \sum_{I_* \vee I'_* = I} c_0^{I'_1}\cdots c_0^{I'_\beta} \tag{5.32}$$

where $I_* \vee I'_*$ is the finest partition of I having both $I_* = I_1,\ldots,I_\alpha$ and $I'_* = I'_1,\ldots,I'_\beta$ as subpartitions. The sum in (5.32) is thus over all I'_* such that $I_* \vee I'_*$ equals the unpartitioned I.

A partition I'_* such that $I_* \vee I'_* = I$ is sometimes referred to as a complementary partition to I_*. McCullagh (1987) provides, in an appendix, tables of complementary partitions for $|I| \leqslant 8$, determined by a computer; see also McCullagh and Wilks (1988). However, for any I_* with $|I| \leqslant 4$ the complementary partitions are simple to find by means of the Hasse diagrams discussed and exhibited in section 5.8.

Written out, some of the first instances of (5.32) are

$$\begin{aligned}
c^{i,jk} &= c^{i,j,k} + c^{i,j}c^{k}[2],\\
c^{ij,kl} &= c^{i,j,k,l} + c^{i}c^{j,k,l}[4] + c^{i,k}c^{j,l}[2] + c^{i}c^{k}c^{j,l}[4],\\
c^{i,jkl} &= c^{i,j,k,l} + c^{j}c^{i,k,l}[3] + c^{i,j}c^{k,l}[3] + c^{j}c^{k}c^{i,l}[3],\\
c^{i,j,kl} &= c^{i,j,k,l} + c^{k}c^{i,j,l}[2] + c^{i,k}c^{j,l}[2].
\end{aligned} \tag{5.33}$$

In conclusion, we note that the above considerations imply that if $p_1(t)$ and $p_0(t)$ are arbitrary functions related by

$$p_0(t) = \log p_1(t)$$

then, letting again $I = i_1\cdots i_\nu$,

$$p_1^{(I)}(t) = p_1(t)\sum_{\alpha=1}^{|I|}\sum_{I/\alpha} p_0^{(I_1)}(t)\cdots p_0^{(I_\alpha)}(t), \tag{5.34}$$

where $p^{(I)}(t)$, for any function $p(t)$, denotes the partial derivatives of p with respect to $t_{i_1},\ldots,t_{i_\nu}$. On the other hand, we find, by (5.28),

$$p_0^{(I)}(t) = \sum_{\alpha=1}^{|I|}(-1)^{\alpha-1}(\alpha-1)!\,p_1(t)^{-\alpha}\sum_{I/\alpha} p_1^{(I_1)}(t)\cdots p_1^{(I_\alpha)}(t). \tag{5.35}$$

5.5 Cumulants and moments

For an m-dimensional random variate $X=(X^1,\dots,X^m)$ the Taylor expansions around 0 of the moment generating function $M(t)$ and the cumulant generating function $K(t)$ may be written as

$$M(t)=1+\sum_{\nu=1}^{\infty}\frac{1}{\nu!}\kappa^{i_1\cdots i_\nu}t_{i_1}\cdots t_{i_\nu} \tag{5.36}$$

and

$$K(t)=\sum_{\nu=1}^{\infty}\frac{1}{\nu!}\kappa^{i_1,\dots,i_\nu}t_{i_1}\cdots t_{i_\nu}, \tag{5.37}$$

where

$$\kappa^{i_1\cdots i_\nu}=E(X^{i_1}\cdots X^{i_\nu}), \tag{5.38}$$

$$\kappa^{i_1,\dots,i_\nu}=K\{X^{i_1},\dots,X^{i_\nu}\} \tag{5.39}$$

are, respectively, the joint moment and the joint cumulant of $X^{i_1},\dots,X^{i_\nu}$.

Note that the notation in (5.38) and (5.39) is different from that employed in section 5.1. Thus we have, for example $E(X^1X^2)=\kappa^{12}$, rather than μ'_{12}, and $\operatorname{cov}(X^1,X^2)=\kappa^{1,2}$, rather than κ_{12}. For many purposes the present notation, as further developed below, is simpler and more transparent to work with.

Cumulants have the following multilinearity property. Let

$$Y^1=a^1_{i_1}X^{i_1}_1,\dots,Y^m=a^m_{i_m}X^{i_m}_m$$

be m linear combinations Y^r of random variables X^i_r ($r=1,\dots,m$; $i=1,2,\dots,i_r$), the a^r_i being nonrandom coefficients. (Note that here the range of the index i may depend on r.) The joint cumulant of $Y^1,\dots,Y^m$ then satisfies the multilinearity relation

$$K\{Y^1,\dots,Y^m\}=a^1_{i_1}\cdots a^m_{i_m}K\{X^{i_1}_1,\dots,X^{i_m}_m\}. \tag{5.40}$$

Example 5.3 Mean linear models

Suppose $Y^1,\dots,Y^m$ are random variables with mean value structure $E(Y^i)=\beta^r x^i_r$, where x^i_r is a matrix of known covariates and β^r, $r=1,\dots,d$, are unknown parameters. Denoting the cumulants of $Y^1,\dots,Y^m$ by κ^I and letting

$$\beta_{r,s}=x^i_r x^j_s\kappa_{i,j}$$

where $\kappa_{i,j}$, here assumed known, is the precision matrix, we have that the weighted least squares estimate b^r of β^r is given by

$$\beta_{r,s}b^s = x_r^i \kappa_{i,j} Y^j$$

or, writing $Y_i = \kappa_{i,j} Y^j$, $b_r = \beta_{r,s}b^s$ and $\beta^{r,s}$ for the inverse matrix of $\beta_{r,s}$,

$$b^r = \beta^{r,s} x_s^i Y_i.$$

By (5.40) the cumulants of b^r, which we denote by $\beta^{r_1,\ldots,r_v}$, are

$$\beta^{r_1,\ldots,r_v} = \beta^{r_1,s_1} \cdots \beta^{r_v,s_v} x_{s_1}^{i_1} \cdots x_{s_v}^{i_v} \kappa_{i_1,\ldots,i_v},$$

where

$$\kappa_{i_1,\ldots,i_v} = \kappa_{i_1,j_1} \cdots \kappa_{i_v,j_v} \kappa^{j_1,\ldots,j_v}.$$

Note that $\beta^{r_1,\ldots,r_v}$ is a contravariant Cartesian tensor relative to linear transformations of the parameter $\beta = (\beta^1, \ldots, \beta^d)$.

As an alternative notation for (5.38) and (5.39) we shall sometimes write

$$\kappa_1^I = \kappa_1^{i_1 \cdots i_v} = E(X^{i_1} \cdots X^{i_v}) \tag{5.41}$$

and

$$\kappa_0^I = \kappa_0^{i_1 \cdots i_v} = K\{X^{i_1}, \ldots, X^{i_v}\}, \tag{5.42}$$

where $I = i_1 \cdots i_v$. Then, as special cases of the exlog relations (5.27) and (5.28) discussed in section 5.4, we obtain the formulae expressing moments in terms of cumulants:

$$\kappa_1^I = \sum_{\alpha=1}^{|I|} \sum_{I/\alpha} \kappa_0^{I_1} \cdots \kappa_0^{I_\alpha} \tag{5.43}$$

and, conversely, cumulants in terms of moments:

$$\kappa_0^I = \sum_{\alpha=1}^{|I|} (-1)^{\alpha-1}(\alpha-1)! \sum_{I/\alpha} \kappa_1^{I_1} \cdots \kappa_1^{I_\alpha}. \tag{5.44}$$

Here, as in section 5.4, the inner sums are over all partitions $I_1, \ldots, I_\alpha$ of $I = i_1 \cdots i_v$ into α blocks.

In particular, since $\kappa_1^I = \kappa^{i_1 \cdots i_v}$ and $\kappa_0^I = \kappa^{i_1,\ldots,i_v}$,

$$\begin{aligned}
\kappa^{ij} &= \kappa^{i,j} + \kappa^i \kappa^j, \\
\kappa^{ijk} &= \kappa^{i,j,k} + \kappa^i \kappa^{j,k}[3] + \kappa^i \kappa^j \kappa^k, \\
\kappa^{ijkl} &= \kappa^{i,j,k,l} + \kappa^i \kappa^{jkl}[4] + \kappa^{i,j}\kappa^{k,l}[3] + \kappa^i \kappa^j \kappa^{k,l}[6] + \kappa^i \kappa^j \kappa^k \kappa^l,
\end{aligned} \tag{5.45}$$

these relations being, in essence, just restatements of the equations

(5.29). The inverse relations corresponding to (5.45) are

$$\begin{aligned}\kappa^{i,j} &= \kappa^{ij} - \kappa^i\kappa^j,\\ \kappa^{i,j,k} &= \kappa^{ijk} - \kappa^i\kappa^{jk}[3] + 2\kappa^i\kappa^j\kappa^k,\\ \kappa^{i,j,k,l} &= \kappa^{ijkl} - \kappa^i\kappa^{jkl}[4] - \kappa^{ij}\kappa^{kl}[3] + 2\kappa^i\kappa^j\kappa^{kl}[6] - 6\kappa^i\kappa^j\kappa^k\kappa^l,\end{aligned} \tag{5.46}$$

cf. (5.30). Note that the second of the above relations shows that

$$\kappa^{i,j,k} = E\{(X^i - \kappa^i)(X^j - \kappa^j)(X^k - \kappa^k)\},$$

i.e. the third-order cumulants as well as those of the second order, equal the corresponding centralized moments. This does not generalize to the higher-order cumulants. Further, we observe that if the mean values κ^i are 0 then

$$\kappa^{i,j,k,l} = \kappa^{ijkl} - \kappa^{i,j}\kappa^{k,l}[3]. \tag{5.47}$$

Taking in particular $i=j=k=l$ we recover the formula $\mu_4 = \kappa_4 + 3\kappa_2^2$ expressing the fourth central moment of a single random variable in terms of its cumulants.

For any subpartition $I_* = I_1, \ldots, I_\alpha$ let us define a *generalized cumulant* by

$$\kappa^{I_*} = \kappa^{I_1,\ldots,I_\alpha} = K\left\{\prod_{i\in I_1} X^i, \ldots, \prod_{i\in I_\alpha} X^i\right\}. \tag{5.48}$$

Note that this notation is consistent with that of (5.38) and (5.39). To distinguish the quantities (5.48) from the special case (5.39) we sometimes refer to the latter as *elemental cumulants*. In particular, we may describe κ^{I_*} as the elemental cumulant of $Y^1, \ldots, Y^\alpha$, where

$$Y^a = \prod_{i\in I_a} X^i, \qquad a = 1, \ldots, \alpha.$$

Example 5.4 Sums of independent random vectors

Let S_n be the sum of n independent copies of $X = (X^1, \ldots, X^m)$ and, as usual, let $S_n^* = \{S_n - E(S_n)\}/\sqrt{n}$. Denoting generalized cumulants of X and of S_n^* by κ^{I_*} and κ^{*I_*}, respectively, we have, using $K(S_n;t) = nK(X;t)$ and (5.40), for $|I| > 1$

$$\kappa^{*I_*} = n^{-1/2|I|+1}\kappa^{I_*}. \tag{5.49}$$

Comparing (5.44) and (5.48), we find that κ^{I_*} can be expressed in terms of moments by

$$\kappa^{I*} = \sum_{I_* \leqslant I'_*} (-1)^{\beta-1}(\beta-1)!\kappa_1^{I'_1}\cdots\kappa_1^{I'_\beta}, \tag{5.50}$$

the summation being over all partitions $I'_* = I'_1, \ldots, I'_\beta$ of I into β blocks, such that $I_* = I_1, \ldots, I_\alpha$ is a subpartition of I'_*. Further, on comparing (5.50) to (5.31), one sees that the generalized cumulants are, in fact, a special case of the generalized c-coefficients defined by (5.31) and hence that the generalized cumulants may be expressed in terms of elemental cumulants by

$$\kappa^{I_*} = \sum_{I_* \vee I'_* = I} \kappa_0^{I'_1}\cdots\kappa_0^{I'_\beta}, \tag{5.51}$$

where $I_* = I_1, \ldots, I_\alpha$ and the sum is over all partitions $I'_* = I'_1, \ldots, I'_\beta$ such that there exists no partition of I, into two or more blocks, having both I_* and I'_* as subpartitions. Thus, for instance, $\kappa^{i,jk} = \kappa^{i,j,k} + \kappa^{i,j}\kappa^k[2]$, this and the corresponding formulae for $\kappa^{ij,kl}$ and $\kappa^{i,jkl}$ being just applications of the relations (5.33).

Example 5.5 Multivariate normal distribution

Let $X = (X^1, \ldots, X^m)$ follow the multivariate normal distribution $\mathrm{MN}_m(\xi, \Sigma)$. The moment generating function of X is

$$M(t) = \exp(t\cdot\xi - \tfrac{1}{2}t\Sigma t^{\mathrm{T}}) \tag{5.52}$$

and hence only the cumulants of order 1 and 2 are different from 0, and these are given by $\kappa^i = \xi_i$ and $\kappa^{i,j} = \sigma_{ij}$, where ξ_i and σ_{ij} are the elements of ξ and Σ, respectively. To determine the central moments of X we assume, without loss of generality, that $\xi = 0$ and using (5.43) we find that all central moments of odd order are 0 and that the first even-order central moments are

$$\kappa^{ij} = \sigma_{ij}, \quad \kappa^{ijkl} = \sigma_{ij}\sigma_{kl}[3], \quad \kappa^{ijklmn} = \sigma_{ij}\sigma_{kl}\sigma_{mn}[15]. \tag{5.53}$$

In the case of a bivariate normal distribution we thus have, for example, $\kappa^{1112} = 3\sigma_{11}^3\sigma_{22}\rho_{12}$, in the previous notation.

Example 5.6 Bivariate log normal distribution (continued)

In Example 5.1 we recorded some of the simpler properties of the bivariate log normal distribution. Note that in the new notation we have, for example, $Y = (Y^1, Y^2)$ and

$$\begin{aligned}
\kappa^1 &= \exp(\mu_1 + \tfrac{1}{2}\sigma_{20}), \qquad \kappa^{1,1} = \exp(2\mu_1 + \sigma_{20})(e^{\sigma_{20}} - 1), \\
\kappa^{1,2} &= \exp(\mu_1 + \mu_2 + \tfrac{1}{2}\sigma_{20} + \tfrac{1}{2}\sigma_{02})(e^{\sigma_{11}} - 1), \\
\kappa^{1,1,1} &= \exp(3\mu_1 + \tfrac{3}{2}\sigma_{20})(e^{\sigma_{20}} - 1)^2(e^{\sigma_{20}} + 2), \\
\kappa^{1,1,2} &= \exp(2\mu_1 + \mu_2 + \sigma_{20} + \tfrac{1}{2}\sigma_{02})(e^{\sigma_{11}} - 1)(e^{\sigma_{20}+\sigma_{11}} + e^{\sigma_{20}} - 2).
\end{aligned} \tag{5.54}$$

Corresponding to the generalized cumulants κ^{I_*} of $X^1, \ldots, X^m$ we define arrays κ_{I_*} by lowering the indices using the precision $\kappa_{i,j}$, i.e.

$$\kappa_{I_1,\ldots,I_\alpha} = \kappa_{i_1,j_1} \cdots \kappa_{i_\nu,j_\nu} \kappa^{J_1,\ldots,J_\alpha},$$

where $I = i_1 \cdots i_\nu$, $J = j_1 \cdots j_\nu$ and it is to be understood that $|J_1| = |I_1|, \ldots, |J_\alpha| = |I_\alpha|$.

5.6 Cumulants of power series

Let $U = (U^1, \ldots, U^d)$ be a d-dimensional random variate determined as a power series in $X = (X^1, \ldots, X^m)$, i.e. denoting generic coordinates of U and X by $U^r, U^s, \ldots$ and $X^i, X^j, \ldots$, respectively, we have

$$U^r = c_0^r + \sum_{|I|=1}^{\infty} c_I^r X^I, \tag{5.55}$$

where the c's are constant coefficients, I is an index set $i_1 \cdots i_\nu$, and $X^I = X^{i_1} \cdots X^{i_\nu}$. The power series need not, of course, be infinite. We wish to express the cumulants of U in terms of the cumulants of X.

Suppose for simplicity that the constant terms c_0^r are 0. Elementary, though somewhat lengthy, calculations (for details, see McCullagh, 1987) show that the elemental cumulants of U are given by

$$\kappa^{r_1,\ldots,r_\nu} = \sum_{|I|=\nu}^{\infty} \sum_{I/\nu} c_{I_1}^{r_1} \cdots c_{I_\nu}^{r_\nu} \kappa^{I_1,\ldots,I_\nu}, \tag{5.56}$$

where the inner summation is over all partitions into ν (nonempty) blocks of all index sets I of length $|I|$.

Thus, in particular and whether c_0^r is 0 or not, we have

$$\begin{aligned}
\kappa^r &= c_0^r + c_i^r \kappa^i + c_{i_1 i_2}^r \kappa^{i_1 i_2} + c_{i_1 i_2 i_3}^r \kappa^{i_1 i_2 i_3} + \cdots, \\
\kappa^{r_1,r_2} &= c_{i_1}^{r_1} c_{i_2}^{r_2} \kappa^{i_1,i_2} + c_{i_1}^{r_1} c_{i_2 i_3}^{r_2} \kappa^{i_1,i_2 i_3}[2] + c_{i_1 i_2}^{r_1} c_{i_3 i_4}^{r_2} \kappa^{i_1 i_2,i_3 i_4} + \cdots, \\
\kappa^{r_1,r_2,r_3} &= c_{i_1}^{r_1} c_{i_2}^{r_2} c_{i_3}^{r_3} \kappa^{i_1,i_2,i_3} + c_{i_1}^{r_1} c_{i_2}^{r_2} c_{i_3 i_4}^{r_3} \kappa^{i_1,i_2,i_3 i_4}[3] \\
&\quad + c_{i_1}^{r_1} c_{i_2 i_3}^{r_2} c_{i_4 i_5}^{r_3} \kappa^{i_1,i_2 i_3,i_4 i_5}[3] + \cdots, \\
\kappa^{r_1,r_2,r_3,r_4} &= c_{i_1}^{r_1} c_{i_2}^{r_2} c_{i_3}^{r_3} c_{i_4}^{r_4} \kappa^{i_1,i_2,i_3,i_4} + c_{i_1}^{r_1} c_{i_2}^{r_2} c_{i_3}^{r_3} c_{i_4 i_5}^{r_4} \kappa^{i_1,i_2,i_3,i_4 i_5}[4] + \cdots.
\end{aligned} \tag{5.57}$$

The formulae relating moments of U to moments of X are considerably more complicated and will not be given here.

It follows immediately from (5.56) that in the repeated sampling case, where $S_n^* = n^{-1/2}(X_1 + \cdots + X_n - n\mu)$, with $\mu = E(X_1)$ and where (5.49) holds, we have on letting $X = S_n^*$ in (5.56) that the asymptotic order of $\kappa^{r_1,\ldots,r_v}$ is $O(n^{-v/2+1})$. More precisely, (5.56) provides an asymptotic expansion of $\kappa^{r_1,\ldots,r_v}$ in powers of $n^{-1/2}$, the leading term of which is $c_{i_1}^{r_1} \cdots c_{i_v}^{r_v} \kappa^{i_1,\ldots,i_v}$. This expansion is, in fact, a valid asymptotic expansion provided (5.55), with $X = S_n^*$, is valid as an asymptotic expansion (rather than being a convergent power series). It follows, in particular, that if for some calculation we are only interested in cumulants or moments of U of asymptotic order $O(n^{-v/2+1})$ or less, then we need take into account only cumulants or moments of X_1 of rank v or less.

These statements require some mild regularity conditions of which generally, from the viewpoint of statistical applications, the only important one is that of existence of the relevant cumulants or moments.

Example 5.7 Quadratic form in multivariate normal random variables

Suppose $(X^1, \ldots, X^m)$ follows the m-dimensional normal distribution with mean 0 and covariance matrix $\lambda^{i,j}$. For a quadratic form

$$U = c_{ij} X^i X^j$$

the rth cumulant of U, denoted in the power notation by $\kappa_r(U)$, is

$$\kappa_r(U) = c_{i_1 j_1} \cdots c_{i_r j_r} \lambda^{i_1, j_2} \lambda^{i_2, j_3} \ldots \lambda^{i_{r-1}, j_r} \lambda^{i_r, j_1} [2^{r-1}(r-1)!] \qquad (5.58)$$

as follows simply from formula (5.56), on noting that the only terms of (5.56) which can contribute to $\kappa_r(U)$ are those for which $|I| = 2r$ and all blocks of the partition $I/(2r)$ have size 2.

5.7 Appendix: Tensorial Hermite polynomials

In section 1.6 we defined univariate orthogonal polynomials and in particular introduced Hermite polynomials associated with the standardized normal distribution. Similar general definitions apply to orthogonal polynomials associated with an arbitrary multivariate distribution but to exploit the preservation of multivariate normality under nonsingular linear transformations definitions and discussion in tensorial notation are natural. As usual we denote the coordinates by $x^1, \ldots, x^m$ and use the summation convention.

Note first that quite generally if the defining distribution has independent components, a system of orthogonal polynomials for the multivariate distribution is formed by taking all possible products of the polynomials associated with the marginal distributions. In particular for the bivariate normal distribution of zero mean with unit covariance matrix, $MN_2(0, I)$, the resulting system is

$$1; H_1(x^1), H_1(x^2); H_2(x^1), H_1(x^1)H_1(x^2), H_2(x^2); \ldots \quad (5.59)$$

Orthogonality is easily proved from the independence and there are enough polynomials in the set to represent, via linear combinations, an arbitrary polynomial in $x^1, \ldots, x^m$.

The orthogonal polynomials for a general multivariate normal distribution of zero mean are in effect obtained by linear transformation from (5.59). It is, however, more convenient to work directly under an invariantly phrased definition in terms of derivatives, indeed with a definition quite parallel to that used in the univariate case.

For this, let $\phi_m(x; \kappa)$ be the density of $MN_m(0, \kappa)$, the m-dimensional multivariate normal distribution of zero mean and covariance matrix $\kappa = [\kappa^{i,j}]$, with $\kappa^{i,j} = E(X^i X^j)$. Thus

$$\phi_m(x; \kappa) = (2\pi)^{-1/2m}\{\det(\kappa)\}^{-1/2} \exp(-\tfrac{1}{2}x^i x^j \kappa_{i,j}), \quad (5.60)$$

where the matrix inverse to κ has elements $\kappa_{i,j}$.

The associated covariant Hermite polynomials denoted by $h_{i_1 \cdots i_k}(x; \kappa)$ are defined by

$$\phi_m(x; \kappa) h_{i_1 \cdots i_k}(x; \kappa) = (-1)^k \partial_{i_1} \cdots \partial_{i_k} \phi_m(x; \kappa), \quad (5.61)$$

where $\partial_i = \partial/\partial x^i$. Note the direct parallel with the definition of the standard Hermite polynomials (section 1.6), namely

$$\phi(x) H_k(x) = (-1)^k (d/dx)^k \phi(x).$$

Introducing the notation

$$x_i = \kappa_{i,j} x^j,$$

we may write the first six covariant Hermite polynomials as

$$\begin{aligned}
h_i &= x_i, \\
h_{ij} &= x_i x_j - \kappa_{i,j}, \\
h_{ijk} &= x_i x_j x_k - \kappa_{i,j} x_k [3], \\
h_{ijkl} &= x_i x_j x_k x_l - \kappa_{i,j} x_k x_l [6] + \kappa_{i,j} \kappa_{k,l} [3], \\
h_{ijklm} &= x_i x_j x_k x_l x_m - \kappa_{i,j} x_k x_l x_m [10] + \kappa_{i,j} \kappa_{k,l} x_m [15], \\
h_{ijklmn} &= x_i x_j x_k x_l x_m x_n - \kappa_{i,j} x_k x_l x_m x_n [15] \\
&\quad + \kappa_{i,j} \kappa_{k,l} x_m x_n [45] - \kappa_{i,j} \kappa_{k,l} \kappa_{m,n} [15].
\end{aligned}$$

Note that the two quantities, on the left and right, in (5.61) equal the Fourier transform $\overleftarrow{\chi}_{i_1 \cdots i_k}(x; \kappa)$ of

$$\chi_{i_1 \cdots i_k}(t; \kappa) = t_{i_1} \cdots t_{i_k} \zeta_m(t; \kappa),$$

where $\zeta_m(t; \kappa)$ is the characteristic function of $\phi_m(x; \kappa)$, i.e.

$$h_{i_1 \cdots i_k}(x; \kappa) = \phi_m(x; \kappa)^{-1} \overleftarrow{\chi}_{i_1 \cdots i_k}(x; \kappa).$$

We denote the Fourier transform of a function f by $\overrightarrow{f}$ and the inverse Fourier transform of f by $\overleftarrow{f}$; cf. section 6.11.

If in (5.61) the components are independent and of unit variance, then $\kappa = I$ and we recover the product definition (5.59) with $h_{11}(x; I) = H_2(x^1)$, etc.

An important consequence of the definition (5.61) is that under nonsingular transformation of the variables the polynomials transform simply, behaving as a covariant Cartesian tensor. To see this, suppose that $\tilde{x}^a = c_i^a x^i$ is a transformation to new variables, these having covariance matrix $\tilde{\kappa}$ with $\tilde{\kappa}^{a,b} = c_i^a c_j^b \kappa^{i,j}$. Under such transformations the exponent of the multivariate normal distribution is invariant. We can now apply (5.61) twice, once directly to obtain $h_{i_1 \cdots i_k}(x; \kappa)$ and secondly to obtain $h_{a_1 \cdots a_k}(\tilde{x}; \tilde{\kappa})$. It is easily shown that

$$h_{a_1 \cdots a_k}(\tilde{x}; \tilde{\kappa}) = c_{a_1}^{i_1} \cdots c_{a_k}^{c_k} h_{i_1 \cdots i_k}(x; \kappa), \tag{5.62}$$

which is the required transformation law for a covariant Cartesian tensor. In one dimension the linear transformation $\tilde{x} = \sigma x$ changes $N(0, 1)$ into $N(0, \sigma^2)$ and the orthogonal polynomials associated with the latter are then $\sigma^k H_k(x) = \sigma^k H_k(\tilde{x}/\sigma)$.

The contravariant form of the polynomials is produced by 'raising' suffixes by the covariance matrix to give

$$h^{i_1 \cdots i_k}(x; \kappa) = \kappa^{i_1, j_1} \cdots \kappa^{i_k, j_k} h_{j_1 \cdots j_k}(x; \kappa). \tag{5.63}$$

An equivalent and more direct definition is to write $\partial^i = \kappa^{i,j}\partial_j$ and then to define

$$\phi_m(x;\kappa)h^{i_1\cdots i_k}(x;\kappa) = (-1)^k \partial^{i_1}\cdots\partial^{i_k}\phi_m(x;\kappa). \tag{5.64}$$

The first few contravariant Hermite polynomials are

$$\begin{aligned}
h^i &= x^i,\\
h^{ij} &= x^i x^j - \kappa^{i,j},\\
h^{ijk} &= x^i x^j x^k - \kappa^{i,j}x^k[3],\\
h^{ijkl} &= x^i x^j x^k x^l - \kappa^{i,j}x^k x^l[6] + \kappa^{i,j}\kappa^{k,l}[3],\\
h^{ijklm} &= x^i x^j x^k x^l x^m - \kappa^{i,j}x^k x^l x^m[10] + \kappa^{i,j}\kappa^{k,l}x^m[15],\\
h^{ijklmn} &= x^i x^j x^k x^l x^m x^n - \kappa^{i,j}x^k x^l x^m x^n[15] + \kappa^{i,j}\kappa^{k,l}x^m x^n[45]\\
&\quad - \kappa^{i,j}\kappa^{k,l}\kappa^{m,n}[15]
\end{aligned}$$

and the immediate parallel with section 1.6, the one-dimensional case, is clear. Compare also with the corresponding covariant forms listed previously, which are obtained from the contravariant forms above by lowering of indices. Quite generally, the term in $h^{i_1\cdots i_m}(x;\lambda)$ of highest polynomial degree has coefficient 1. If m is even then every term in $h^{i_1\cdots i_m}$ is of even order and if m is odd then every term in $h^{i_1\cdots i_m}$ has odd order. The general formula for $h^{i_1\cdots i_m}$ is

$$h^{i_1\cdots i_m} = \sum_{\nu=0}^{\langle m/2]} (-1)^\nu \kappa^{i_1,i_2}\cdots\kappa^{i_{2\nu-1},i_{2\nu}}x^{i_{2\nu+1}}\cdots x^{i_m}[\{2^\nu 1^{m-2\nu}\}], \tag{5.65}$$

where $\langle m/2]$ is the smallest integer less than or equal to $\frac{1}{2}m$ and where $\{2^\nu 1^{m-2\nu}\}$ denotes the number of partitions of $\{1,2,\ldots,m\}$ into ν pairs and $m-2\nu$ singletons.

It is useful also to list the inversions of the above formulae for the lower-order contravariant Hermite polynomials:

$$\begin{aligned}
x^0 &= 1,\\
x^i &= h^i,\\
x^i x^j &= h^{ij} + \kappa^{i,j},\\
x^i x^j x^k &= h^{ijk} + \kappa^{i,j}h^k[3],\\
x^i x^j x^k x^l &= h^{ijkl} + \kappa^{i,j}h^{kl}[6] + \kappa^{i,j}\kappa^{k,l}[3],\\
x^i x^j x^k x^l x^m &= h^{ijklm} + \kappa^{i,j}h^{klm}[10] + \kappa^{i,j}\kappa^{k,l}h^m[15],\\
x^i x^j x^k x^l x^m x^n &= h^{ijklmn} + \kappa^{i,j}h^{klmn}[15] + \kappa^{i,j}\kappa^{k,l}h^{mn}[45]\\
&\quad + \kappa^{i,j}\kappa^{k,l}\kappa^{m,n}[15].
\end{aligned}$$

The general formula is

$$x^{i_1}\cdots x^{i_m}=\sum_{\nu=0}^{\langle m/2]}\kappa^{i_1,i_2}\cdots\kappa^{i_{2\nu-1},i_{2\nu}}h^{i_{2\nu+1}}\cdots h^{i_m}[\{2^\nu 1^{m-2\nu}\}]. \tag{5.66}$$

For our particular purposes in connexion with Edgeworth series, in many ways the most important property of the tensorial Hermite polynomials is connected with the expansion of moment generating functions:

$$\int e^{t_jx^j}h_{i_1\cdots i_k}(x;\kappa)\phi_m(x;\kappa)dx=t_{i_1}\cdots t_{i_k}\exp(-\tfrac{1}{2}\kappa^{i,j}t_it_j). \tag{5.67}$$

Note that this formula is immediate from one-dimensional results when κ is the identity matrix and that the general form follows on applying a suitable linear transformation.

Other properties of importance include recursion and orthogonality, i.e.

$$\partial_i h_{i_1\cdots i_k}(x;\kappa)=\kappa_{ii_1}h_{i_2\cdots i_k}(x;\kappa)[k], \tag{5.68}$$

and

$$\int h_{i_1\cdots i_k}(x;\kappa)h_{j_1\cdots j_l}(x;\kappa)\phi(x;\kappa)dx$$

$$=\begin{cases}\kappa_{i_1j_1}\cdots\kappa_{i_kj_k}[k!] & (l=k),\\ 0 & \text{otherwise.}\end{cases} \tag{5.69}$$

To derive the properties (5.68) and (5.69), we write $\phi_m(x+y;\kappa)$ in two ways. First, by a simple rearrangement of terms

$$\phi(x+y;\kappa)=\exp(\kappa_{i,j}y^ix^j-\tfrac{1}{2}\kappa_{i,j}y^iy^j)\phi_m(x;\kappa).$$

On the other hand, we have by Taylor's formula,

$$\phi_m(x+y;\kappa)=\sum_{\nu=0}^{\infty}\frac{1}{\nu!}h_{i_1\cdots i_\nu}(x;\kappa)y^{i_1}\cdots y^{i_\nu}\phi_m(x;\kappa).$$

Comparing these two formulae, we find

$$\exp(\kappa_{i,j}y^ix^j-\tfrac{1}{2}\kappa_{i,j}y^iy^j)=\sum_{\nu=0}^{\infty}\frac{1}{\nu!}h_{i_1\cdots i_\nu}(x;\kappa)y^{i_1}\cdots y^{i_\nu} \tag{5.70}$$

and hence $h_{i_1\cdots i_\nu}$ may be determined as the coefficient of $y^{i_1}\cdots y^{i_\nu}/\nu!$ in the power series expansion in the y^i around 0 of the left-hand side of (5.70). Now, differentiating (5.70) with respect to x and identifying coefficients we obtain (5.68), and (5.69) may be proved by integration by parts, using (5.68).

Formula (5.61) shows how the tensorial Hermite polynomials $h_{i_1 \cdots i_k}$ are related to the derivatives of $\phi_m(x;\kappa)$ with respect to the components of x. For k even, the polynomials are also related to the derivatives of $\phi(x;\kappa)$ with respect to the components of the covariance matrix $\kappa^{i,j}$. Specifically, letting $\tilde{\partial}_{ij} = \partial/\partial\kappa^{i,j}$, we have for $p = 1, 2, \ldots$

$$\tilde{\partial}_{i_1 j_1} \cdots \tilde{\partial}_{i_p j_p} \phi_m(x;\kappa) = 2^{-(\delta_{i_1 j_1} + \cdots + \delta_{i_p j_p})} h_{i_1 j_1 \cdots i_p j_p}(x;\kappa)\phi_m(x;\kappa), \quad (5.71)$$

where δ_{ij} is the Kronecker delta. Comparing this to (5.61), we find that (5.71) is equivalent to

$$\tilde{\partial}_{i_1 j_1} \cdots \tilde{\partial}_{i_p j_p} \phi_m(x;\kappa) = 2^{-(\delta_{i_1 j_1} + \cdots + \delta_{i_p j_p})} \partial_{i_1} \partial_{j_1} \cdots \partial_{i_p} \partial_{j_p} \phi_m(x;\kappa). \quad (5.72)$$

Formulae (5.71) and (5.72) may be proved by induction. For $p = 1$ a short direct calculation shows that

$$\frac{\partial \phi_m(x;\kappa)}{\partial \kappa^{i,j}} = 2^{-\delta_{ij}} \frac{\partial^2 \phi_m(x;\kappa)}{\partial x^i \partial x^j}, \quad (5.73)$$

i.e. in this case that (5.72) holds, and using this and (5.61) the step from p to $p+1$ follows from

$$\begin{aligned}
&\tilde{\partial}_{i_1 j_1} \cdots \tilde{\partial}_{i_{p+1} j_{p+1}} \phi_m(x;\kappa) \\
&= \tilde{\partial}_{i_{p+1} j_{p+1}} \{2^{-(\delta_{i_1 j_1} + \cdots + \delta_{i_p j_p})} h_{i_1 j_1 \cdots i_p j_p}(x;\kappa)\phi_m(x;\kappa)\} \\
&= 2^{-(\delta_{i_1 j_1} + \cdots + \delta_{i_p j_p})} [\phi_m(x;\kappa) \tilde{\partial}_{i_{p+1} j_{p+1}} \{\phi_m(x;\kappa)^{-1} \partial_{i_1} \cdots \partial_{j_p} \phi_m(x;\kappa)\} \\
&\quad + h_{i_1 j_1 \cdots i_p j_p}(x;\kappa) 2^{-\delta_{i_{p+1} j_{p+1}}} h_{i_{p+1} j_{p+1}}(x;\kappa)\phi_m(x;\kappa)] \\
&= 2^{-(\delta_{i_1 j_1} + \cdots + \delta_{i_{p+1} j_{p+1}})} [-h_{i_{p+1} j_{p+1}}(x;\kappa) \partial_{i_1} \cdots \partial_{j_p} \phi_m(x;\kappa) \\
&\quad + \partial_{i_1} \cdots \partial_{j_{p+1}} \phi_m(x;\kappa) + h_{i_1 j_1 \cdots i_p j_p}(x;\kappa) h_{i_{p+1} j_{p+1}}(x;\kappa)\phi_m(x;\kappa)] \\
&= 2^{-(\delta_{i_1 j_1} + \cdots + \delta_{i_{p+1} j_{p+1}})} \partial_{i_1} \cdots \partial_{j_{p+1}} \phi_m(x;\kappa).
\end{aligned}$$

For important applications of (5.73) in the study of level upcrossings of Gaussian processes, see Slepian (1962) and Cramér and Leadbetter (1967, p. 268). Cf. also Further results and exercises 5.9 and 5.10, and Plackett (1954).

5.8 Appendix: Partially ordered sets, partitions and Möbius inversion

Let $\mathscr{P}$ be a set whose points we denote by ρ, σ, τ, etc. Then $\mathscr{P}$ is said to be *partially ordered* if for any pair of points ρ and σ either there is no relation between these points or $\rho \leqslant \sigma$ or $\sigma \leqslant \rho$, where $\leqslant$, to be read 'precedes', indicates an abstract relation which is transitive ($\rho \leqslant \sigma$ and $\sigma \leqslant \tau$ implies $\rho \leqslant \tau$), reflexive ($\rho \leqslant \rho$) and such that if $\rho \leqslant \sigma$ and $\sigma \leqslant \rho$ then $\rho = \sigma$, i.e. ρ and σ are identical. We write $\rho < \sigma$ if $\rho \leqslant \sigma$ but ρ and σ are not identical.

A *partition* of a set E is a collection of disjoint nonempty subsets of E such that the union of these subsets equals E. We also speak of such partitions as set partitions to distinguish them from another concept that we shall touch upon later, that of partitions of integers. The class of all partitions of E is denoted by $\mathscr{P}(E)$.

Example 5.8 Partitions

Let m be a positive integer and write I for the set $\{1, 2, \ldots, m\}$. Letting I_* and I'_* denote elements of $\mathscr{P}(I)$, we define a partial ordering on $\mathscr{P}(I)$ by letting $I_* < I'_*$ indicate that I_* is 'finer' than I'_* in the sense of being obtainable from I'_* by further partitioning. Thus, for instance, if $m = 4$ and $I_* = 1|24|3$ and $I'_* = 1|234$ we have $I_* < I'_*$.

Given a finite partially ordered set $\mathscr{P}$ there exists a complete ordering of $\mathscr{P}$ which respects the partial ordering in the obvious sense. This is easily seen by induction on the number of elements of the set.

A partially ordered set $\mathscr{P}$ is said to be a *lattice* if any pair (ρ, σ) of points of $\mathscr{P}$ possesses a uniquely determined *least upper bound*, denoted $\rho \vee \sigma$, and a uniquely determined *greatest lower bound*, denoted $\rho \wedge \sigma$, the concepts of upper and lower bounds being defined in the obvious way.

Example 5.9 Partition lattices

The partition sets $\mathscr{P}(I)$, discussed in Example 5.8, are in fact lattices. The greatest lower bound $I_* \wedge I'_*$ of two partitions I_* and I'_* of I is simply the partition whose blocks are the nonempty intersections of blocks of I_* with blocks of I'_*.

The various partitions of I may conveniently be thought of as arranged in a *Hasse diagram* which depicts the partial ordering. The general nature of such diagrams appears from the instances shown in Figure 5.1. By natural convention, where two elements are joined by a line, the element to the left is 'finer' than that to the right.

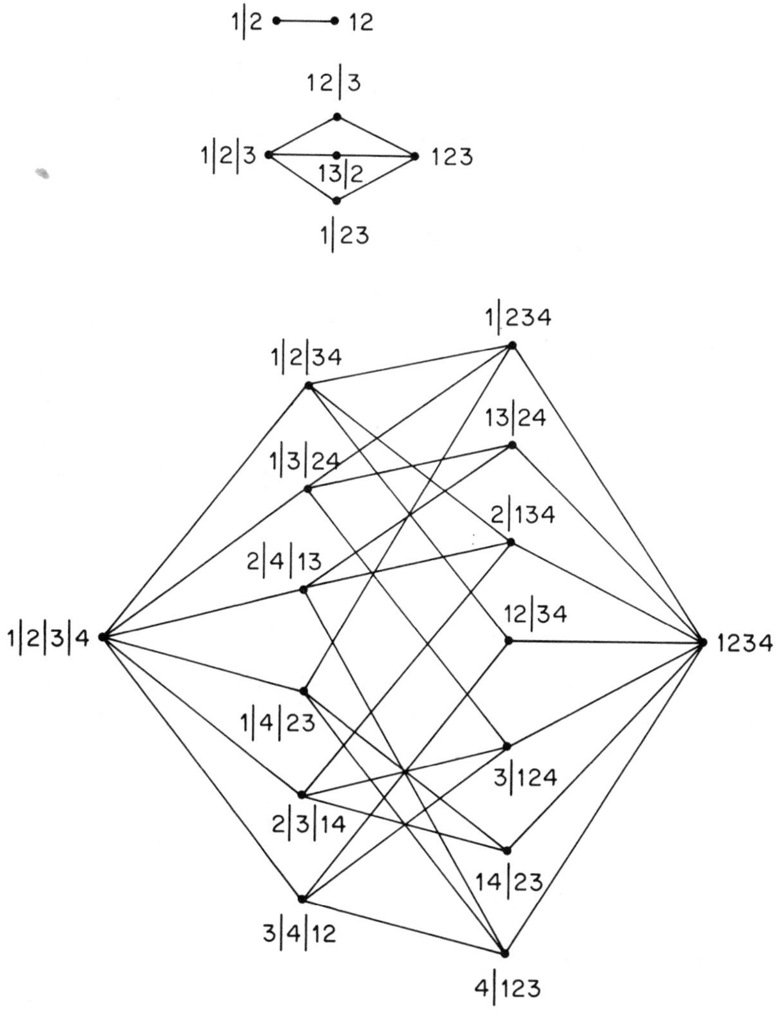

Figure 5.1 *The Hasse diagrams for the partition lattices $\mathscr{P}(I)$, $I = \{1, \ldots, m\}$ and $m = 2, 3, 4$. As an illustration of the lattice operations $\wedge$ and $\vee$ we have* $1|234 \wedge 2|134 = 1|2|34$ *and* $1|2|34 \vee 3|124 = 1234$.

Associated with any finite partially ordered set $\mathscr{P}$ are two numerical functions, both defined on $\mathscr{P} \times \mathscr{P}$: the *Riemann function* or *zeta function* given by

$$\zeta(\rho, \sigma) = \begin{cases} 1 & \text{if } \rho \leqslant \sigma, \\ 0 & \text{otherwise,} \end{cases} \tag{5.74}$$

and the *Möbius function* μ defined as the inverse of ζ in the sense of satisfying

$$\sum_{\sigma} \zeta(\rho, \sigma)\mu(\sigma, \tau) = \delta(\rho, \tau) = \sum_{\sigma} \mu(\rho, \sigma)\zeta(\sigma, \tau), \tag{5.75}$$

where $\delta(\rho, \tau) = 0$ or 1 according as $\rho \neq \tau$ or $\rho = \tau$. Another way to express (5.75) is to say that the matrices $Z = [\zeta(\rho, \sigma)]$ and $M = [\mu(\rho, \sigma)]$, where ρ and σ respectively index rows and columns, are mutually inverse. That ζ is invertible, and μ consequently well defined, follows from (5.74) and the fact that there is a complete ordering of $\mathscr{P}$ which respects the given partial order, for this means that Z can be arranged as an upper triangular matrix with ones along the diagonal.

As we now prove, the Möbius function may alternatively be characterized as being determined recursively by

$$\mu(\rho, \tau) = \begin{cases} 1 & \text{if } \rho = \tau, \\ -\sum\limits_{\rho \leqslant \sigma < \tau} \mu(\rho, \sigma) & \text{if } \rho < \tau, \\ 0 & \text{otherwise.} \end{cases} \tag{5.76}$$

Brief reflection shows that (5.76) does, in fact, uniquely define a function μ and that if $\rho < \tau$ and if there is no σ with $\rho < \sigma < \tau$ then

$$\mu(\rho, \tau) = -\mu(\rho, \rho) = -1.$$

Therefore it suffices to verify that μ given by (5.76) satisfies the second equality, say, in (5.75). By (5.74) and (5.76),

$$\sum_{\sigma} \mu(\rho, \sigma)\zeta(\sigma, \tau) = \sum_{\rho \leqslant \sigma \leqslant \tau} \mu(\rho, \tau)$$

and on account of (5.76) this, indeed, equals $\delta(\rho, \tau)$.

Example 5.10 A simple example

Using Figure 5.1 and formula (5.76) we find that for $\mathscr{P}(123)$ we have $\mu(1|2|3, 123) = 2$ while all other μ values are either $+1$ or -1, while for $\mathscr{P}(1234)$ one finds, for instance, $\mu(12|3|4, 1234) = 2$ and $\mu(1|2|3|4, 1234) = -6$.

Now, let f be any real-valued function on $\mathscr{P}$ and define F on $\mathscr{P}$ by

$$F(\sigma) = \sum_{\rho \leqslant \sigma} f(\rho). \tag{5.77}$$

Thinking of f and F as row vectors, we may write this relation as $F = fZ$ and the inverse relation to (5.74) is therefore

$$f(\sigma) = \sum_{\rho \leqslant \sigma} \mu(\rho, \sigma) F(\rho). \tag{5.78}$$

The step from (5.77) to (5.78) is called *Möbius inversion*. The need for this operation arises in a variety of mathematical and statistical contexts. We shall use it in section 5.9 to prove the important relation (5.32) and some further instances of statistical relevance are indicated in Exercises 5.11 and 5.15. For further insight into the mathematical aspects, see Rota (1964), Aigner (1979), or Stanley (1986).

On any partially ordered set $\mathscr{P}$ with order relation $\leqslant$ we may define a reverse relation $\lesssim$ by the condition that $\rho \lesssim \sigma$ is equivalent to $\sigma \leqslant \rho$. This relation also determines a partial ordering, and under a matrix interpretation, as above, the zeta and Möbius functions of $\lesssim$ correspond to the matrix transposition of the zeta and Möbius functions of $\leqslant$.

It follows that in addition to (5.77)–(5.78) we have that if f and F are arbitrary real functions on $\mathscr{P}$ related by

$$F(\sigma) = \sum_{\sigma \leqslant \rho} f(\rho), \tag{5.79}$$

then

$$f(\sigma) = \sum_{\sigma \leqslant \rho} \mu(\sigma, \rho) F(\rho), \tag{5.80}$$

and vice versa.

In many cases an explicit formula for μ can be found fairly easily.

For instance, as a useful general result we have that for any index set I and any $I_* \in \mathscr{P}(I)$

$$\mu(I_*, I) = (-1)^{|I_*|-1}(|I_*| - 1)!, \tag{5.81}$$

where $|I_*|$ denotes the number of blocks in I_*.

To prove this, recall that any mapping ϕ from I into some other set X determines a partition I_* of I, the blocks of this partition being the constancy sets of ϕ. Now suppose that X is a finite set with $x = |X|$ elements, and given a partition I_* of I let $f(I_*)$ be the number of distinct mappings $\phi: I \to X$ which partition I exactly into I_*. Then, denoting the complete partition of I into $|I|$ singletons by I_0, we have

$$f(I_0) = x(x-1)\cdots(x-|I|+1). \tag{5.82}$$

Furthermore,

$$\sum_{I_* \leqslant I'_*} f(I'_*) = x^{|I_*|}$$

and hence, by (5.80),

$$f(I_*) = \sum_{I_* \leqslant I'_*} \mu(I_*, I'_*) x^{|I'_*|}.$$

In particular, using (5.82), we have

$$x(x-1)\cdots(x-m+1) = \sum \mu(I_0, I_*) x^{|I_*|},$$

where $m = |I|$. Comparing coefficients we find $\mu(I_0, I) = (-1)^{m-1}(m-1)!$, and hence (5.81) is verified for $I_* = I_0$. The general case follows on observing that the restriction of $\mu(I'_*, I''_*)$ to those I'_*, I''_* for which $I_* \leqslant I'_*$ and $I_* \leqslant I''_*$ must be equivalent to the Möbius function for the partition lattice with $|I_*|$ elements.

5.9 Appendix: Proof of the exlog relations

In our discussion of the exlog relations in section 5.4, the verification of (5.32) was left to the present appendix. We now give a proof using Möbius inversion. First, however, we prove (5.27) and (5.28), the basic exlog relations, by a method different to that of section 5.4. For this we also use Möbius inversion.

For the proof we shall need the following, easily verified, combinatorial result.

Lemma 5.1

Let N, ν, $n_1, \ldots, n_\nu$, $N_1, \ldots, N_\nu$ be positive integers all different and such that

$$n_1 N_1 + \cdots + n_\nu N_\nu = N. \tag{5.83}$$

We write $n_* = (n_1, \ldots, n_\nu)$ and $N_* = (N_1, \ldots, N_\nu)$. Further, let I be a set consisting of N elements.

The number of ways in which I can be partitioned into $n_+ = n_1 + \cdots + n_\nu$ blocks such that n_r blocks are of size N_r, $r = 1, \ldots, \nu$, is given by

$$\binom{N}{n_*; N_*} = \frac{N!}{n_1! \cdots n_\nu! (N_1!)^{n_1} \cdots (N_\nu!)^{n_\nu}}. \tag{5.84}$$

We note in passing that (5.83) constitutes a partition of the natural number N.

To verify (5.27) we first in the assumed relation $p_1(t) = \exp\{p_0(t)\}$ expand the exponential by Taylor's formula and then insert the expression (5.26) for $p_0(t)$, thus obtaining

$$p_1(t) = 1 + \sum_{r=1}^{\infty} \frac{1}{r!} \left\{ \sum_{|I|=1}^{\infty} \frac{1}{|I|!} c_0^I t_I \right\}^r. \tag{5.85}$$

The summand may be re-expressed as

$$\sum_{\nu=1}^{r} \sum_{n_+ = r} \frac{1}{n_1! \cdots n_\nu!} \sum_{|I_1| < \cdots < |I_\nu|} \prod_{s=1}^{\nu} \left\{ \frac{1}{|I_s|!} c_0^{I_s} t_{I_s} \right\}^{n_s}.$$

On inserting this in (5.85) one obtains, after some rearrangement of terms,

$$p_1(t) = 1 + \sum_{|I|=1}^{\infty} \frac{1}{|I|!} \sum{}^* \binom{|I|}{n_*; N_*} c_0^{I_{11}} \cdots c_0^{I_{1n_1}} \cdots c_0^{I_{\nu 1}} \cdots c_0^{I_{\nu n_\nu}} t_I.$$

Here the index sets satisfy $I = I_{11} \cdots I_{\nu n_\nu}$ and $|I_{r1}| = \cdots = |I_{rn_r}| = N_r$, $r = 1, \ldots, \nu$, and $\sum^*$ indicates summation over all ν, $n_1, \ldots, n_\nu$, $N_1, \ldots, N_\nu$ satisfying the conditions of Lemma 5.1 with $N = |I|$. On account of that lemma one may now conclude that

$$p_1(t) = 1 + \sum_{|I|=1}^{\infty} \frac{1}{|I|!} \sum_{\alpha=1}^{|I|} \sum_{I/\alpha} c_0^{I_1} \cdots c_0^{I_\alpha} t_I$$

from which (5.27) follows.

It remains to verify formula (5.32), which, as will be recalled, generalizes (5.28). With $I_* = I_1, \ldots, I_\alpha$, let us introduce the notation

$$c_1^{I_*} = c_1^{I_1} \cdots c_1^{I_\alpha}, \qquad c_0^{I_*} = c_0^{I_1} \cdots c_0^{I_\alpha}.$$

Then (5.27) may be written as

$$c_1^I = \sum_{I_* \leqslant I} c_0^{I_*}$$

and multiplying such expressions together we find

$$c_1^{I_*} = \sum_{I'_* \leqslant I_*} c_0^{I'_*}. \tag{5.86}$$

The generalized c-coefficients are defined by (5.31), and in view of (5.81) they may be written as

$$c^{I_*} = \sum_{I_* \leqslant I'_*} \mu(I'_*, I) c_1^{I'_*}.$$

Inserting (5.86) we find

$$\begin{aligned} c^{I_*} &= \sum_{I_* \leqslant I'_*} \sum_{I''_* \leqslant I'_*} \mu(I'_*, I) c_0^{I''_*} \\ &= \sum_{I''_* \leqslant I} c_0^{I''_*} \sum_{I_* \vee I''_* \leqslant I'_*} \mu(I'_*, I) \\ &= \sum_{I''_* \leqslant I} c_0^{I''_*} \sum_{I'_* \leqslant I} \zeta(I_* \vee I''_*, I'_*) \mu(I'_*, I) \\ &= \sum_{I''_* \leqslant I} c_0^{I''_*} \delta(I_* \vee I''_*, I) \\ &= \sum_{I_* \vee I'_* = I} c_0^{I'_*}, \end{aligned}$$

where we have used formula (5.75). This proves (5.32).

Further results and exercises

5.1 Carry through the expansions suggested in Example 5.1 with $\sigma_{ij} = \sigma'_{ij}/t$, $t \to \infty$ and study, in particular, the relation between the correlation coefficients of X and Y. Examine the m-variate log normal distribution and use the techniques of section 5.5 to obtain the third- and fourth-order cumulants.

[Sections 5.1, 5.5]

5.2 Show that elemental cumulants may be represented as

$$K\{X^1, \ldots, X^m\} = \int_{-\infty}^{+\infty} \cdots \int_{-\infty}^{+\infty} K\{\bar{\chi}_{x^1}(X^1), \ldots, \bar{\chi}_{x^m}(X^m)\} \, dx^1 \cdots dx^m, \tag{5.87}$$

where $\bar{\chi}_{x^i}(X^i) = 1$ for $X^i > x^i$, and 0 otherwise. Express the integrand of (5.87) in terms of survival functions.

Show also that if the random variables X^i are nonnegative then

$$E(X^1 \cdots X^m) = \int_0^\infty \cdots \int_0^\infty \bar{F}(x^1, \ldots, x^m) dx^1 \cdots dx^m, \tag{5.88}$$

where $\bar{F}$ is the survival function

$$\bar{F}(x^1, \ldots, x^m) = P(X^1 > x^1, \ldots, X^m > x^m).$$

Generalize the result to arbitrary joint moments.

[Section 5.5; Block and Fang, 1988]

5.3 *An alternative definition of generalized cumulants.* Speed (1983) defined a notion of generalized cumulants which is somewhat different from that defined in section 5.5. Specifically, Speed defines the generalized cumulant of $X^1, \ldots, X^m$ determined by a partition $I_* = I_1, \ldots, I_\alpha$ of $I = 12 \cdots m$ as

$$\kappa_{I_*} = \sum_{I'_*} \mu(I'_*, I_*) \kappa_1^{I_1} \cdots \kappa_1^{I_\alpha}, \tag{5.89}$$

where $\mu(\cdot,\cdot)$ is the Möbius function (5.76), i.e. κ_{I_*} is obtained by Möbius inversion of the 'generalized moments'

$$\kappa_1^{I_*} = \kappa_1^{I_1} \cdots \kappa_1^{I_\alpha}. \tag{5.90}$$

Thus the inverse relation is simply

$$\kappa_1^{I_*} = \sum_{I'_* \leqslant I_*} \kappa_{I'_*}. \tag{5.91}$$

The generalized cumulants κ^{I_*} considered in section 5.5 are related to the alternative κ_{I_*} by

$$\kappa^{I_*} = \sum_{I_* \vee I'_* = I} \kappa_{I'_*}. \tag{5.92}$$

The definition (5.89) has the advantage of generalizing in a natural manner to variance component situations, cf. Speed (1986a, b, c) and Speed and Silcock (1985a, b).

For ordinary (i.e. not generalized) cumulants the two definitions agree as may be seen from (5.92) by setting $I_* = i_1, \ldots, i_\nu$.

[Section 5.5]

5.4 *Sample cumulants.* Sample cumulants, also known as k-statistics, are empirical counterparts of cumulants. To define the sample cumulants, suppose $X_1, \ldots, X_n$ are independent and identically distributed random vectors of dimension m, generic coordinates of these being denoted by X_ν^i, etc. Further, let $\kappa^{I_*} = \kappa^{I_1, \ldots, I_\alpha}$ denote a generalized cumulant of X_ν. For each such cumulant there exists, provided $|I| \leqslant n$, a corresponding k-statistic, i.e. a unique polynomial symmetric function k^{I_*} of the coordinates of $X_1, \ldots, X_n$ such that $E(k^{I_*}) = \kappa^{I_*}$.

Show that

$$\begin{aligned} k^{i,j} &= \{k^{ij} - k^i k^j\} n/(n-1), \\ k^{i,j,k} &= \{k^{ijk} - k^i k^{jk}[3] + 2k^r k^s k^t\} n^2/(n-1)^{(2)}, \\ k^{i,j,k,l} &= \{(n+1)k^{ijkl} - (n+1)k^i k^{jkl}[4] - (n-1)k^{ij}k^{kl}[3] \\ &\quad + 2nk^i k^j k^{kl}[6] - 6nk^i k^j k^k k^l\} n^2/(n-1)^{(3)}, \end{aligned}$$

where $n^{(r)} = n(n-1)\cdots(n-r+1)$.

McCullagh (1987) gives an extensive account of the theory and applicability of sample cumulants.

[Section 5.5]

5.5 *Derived scalars.* From the cumulants and generalized cumulants of a set of random variables $X^1, \ldots, X^m$ it is possible to construct a variety of Cartesian scalars, i.e. real-valued quantities whose values do not change under linear transformations of $X = (X^1, \ldots, X^m)$. Some examples, that all turn up in an interesting way in statistical calculations, are

$$\begin{aligned} \rho_3^2 &= \kappa^{i,j,k}\kappa^{l,p,q}\kappa_{i,j}\kappa_{k,l}\kappa_{p,q}, \\ \tilde{\rho}_3^2 &= \kappa^{i,k,m}\kappa^{j,l,n}\kappa_{i,j}\kappa_{k,l}\kappa_{m,n}, \\ \rho_4 &= \kappa^{i,j,k,l}\kappa_{i,j}\kappa_{k,l}, \\ \tau_4 &= \{\kappa^{i,j,k,l} - \kappa^{ij,kl}[3]\}\kappa_{i,j}\kappa_{k,l}, \\ \tau_{2,2} &= \{k^{ij,kl} + \kappa^{i,j,kl}\}\kappa_{i,j}\kappa_{k,l}. \end{aligned}$$

Note that for $m = 1$ we have that $\rho_3^2 = \tilde{\rho}_3^2$ and that ρ_3 and ρ_4 are, in fact, the ordinary standardized third and fourth cumulants of a one-dimensional random variable.

Calculate the above quantities when $(X^1, \ldots, X^m)$ follow a multinomial distribution with cell probabilities $\pi_1, \ldots, \pi_m$ $(\pi_1 + \cdots + \pi_m = 1)$. Note, in particular, that this shows that ρ_3^2 and $\tilde{\rho}_3^2$ may be different.

[Section 5.5; McCullagh, 1987, section 2.8]

5.6 *Elemental cumulants in terms of conditional cumulants.* For arbitrary random variables $X^1, \ldots, X^m$ and an arbitrary random Y we let $K\{X^1, \ldots, X^m | Y\}$ denote the elemental cumulant of $X^1, \ldots, X^m$ in the conditional distribution given Y. The following formula often allows easy calculation of cumulants via conditioning:

$$K\{X^{i_1}, \ldots, X^{i_\nu}\} = \sum_{\alpha=1}^{\nu} \sum_{I/\alpha} K\{K\{X^i : i \in I_1 | Y\}, \ldots, K\{X^i : i \in I_\alpha | Y\}\}. \tag{5.93}$$

As particular cases, we have the well-known formulae

$$E(X) = E\{E(X | Y)\}, \qquad \mathrm{var}(X) = E\{\mathrm{var}(X | Y)\} + \mathrm{var}\{E(X | Y)\}.$$

[Section 5.5; Brillinger, 1969; Speed, 1983]

5.7 Let U follow a standard normal distribution and define a sequence of random variables X_r, $r = 1, 2, \ldots$, by $X_r = H_r(U)$, where H_r is the rth-order Hermite polynomial. Show that $E(X_r) = 0$ and $\mathrm{var}\,(X_r) = r!$ for $r > 0$, and that the X_r are uncorrelated.

For $\nu = 1, 2, \ldots$ and $r_1, \ldots, r_\nu$ nonnegative integers let

$$c_{r_1 \cdots r_\nu} = E(X_{r_1} \cdots X_{r_\nu})$$

and

$$\bar{c}_{r_1 \cdots r_\nu} = c_{r_1 \cdots r_\nu} / (r_1! \cdots r_\nu!)$$

and let

$$\bar{c}(z_1, \ldots, z_\nu) = \sum_{r_1, \ldots, r_\nu \geqslant 0} \bar{c}_{r_1 \cdots r_\nu} z_1^{r_1} \cdots z_\nu^{r_\nu},$$

i.e. $\bar{c}(z_1, \ldots, z_\nu)$ is the generating function of the coefficients $\bar{c}_{r_1 \cdots r_\nu}$. Show that for $\nu \geqslant 1$

$$\bar{c}(z_1, \ldots, z_\nu) = \exp\left\{ \sum_{1 \leqslant i < j \leqslant \nu} z_i z_j \right\}.$$

[Section 5.7]

5.8 *Wick products and Appell polynomials.* Let $X = (X^1, \ldots, X^m)$ be an m-dimensional random vector with cumulant generating function $k(\theta) = K(X; \theta)$. The tilting factor

$$v(\theta) = V(X; \theta; x) = e^{\theta \cdot x - k(\theta)} \tag{5.94}$$

generating the exponential tilt of the distribution of X, may be viewed as a generating function

$$v(\theta) = \sum_{|I|=0}^{\infty} \frac{1}{|I|!} \theta^I c_I(x), \tag{5.95}$$

the coefficients $c_I(x)$ being polynomials in the coordinates $x^1, \ldots, x^m$ of x with coefficients determined by the cumulants of X.

Show that

$$c_I(x) = \sum_{\alpha=1}^{|I|} \sum_{I/\alpha} (x_{I_1} - \kappa_{I_1}) \cdots (x_{I_\alpha} - \kappa_{I_\alpha}), \tag{5.96}$$

where

$$x_I = \begin{cases} x_i & \text{if } I = i, \\ 0 & \text{otherwise,} \end{cases}$$

and $\kappa_I = k_{/I}(0)$ are the cumulants of X. In particular, if X follows the m-dimensional standard normal distribution, then $c_I(x)$ equals the tensorial Hermite polynomial $h_I(x)$.

Let

$$W\{X^1, \ldots, X^m\} = c_{1\cdots m}(X). \tag{5.97}$$

The random variable $W\{X^1, \ldots, X^m\}$ is called the Wick product of $X^1, \ldots, X^m$. A simple calculation shows that $c_{1\cdots m}(X)$ is equal to the residual of $X^1 \cdots X^m$ after linear regression on all polynomials in $X^1, \ldots, X^m$ of order less than m.

The one-dimensional Appell polynomials corresponding to a random variable Y are defined by assuming that X^i, $i = 1, 2, \ldots$, are all equal to Y and by setting

$$P_n(y) = c_{1\cdots n}(y, \ldots, y) \qquad (y \in R). \tag{5.98}$$

These polynomials satisfy

$$P_n'(y) = nP_{n-1}(y), \quad P_0(y) = 1, \quad E\{P_n(Y)\} = \delta_{0n},$$

where δ is the Kronecker delta.

For $Y \sim \mathrm{N}(0, 1)$ we obtain $P_n = H_n$, the nth-order one-dimensional Hermite polynomial.

Show that if $E(Y) = 0$ and $\mu_n = E(Y^n)$, then

$$y^n = \sum_{k=1}^{n} \binom{n}{k} \mu_{n-k} P_k(y).$$

Appell polynomials may be used to construct noncentral limit

theorems and self-similar processes with long-range dependence, in particular Hermite processes.
[Section 5.7; Taqqu, 1979; Malyshev, 1980; Avram and Taqqu, 1987]

5.9 Let $\phi(x, y; \rho)$ denote the probability density function of the two-dimensional normal distribution with mean $(0,0)$, unit variances and correlation coefficient ρ, and let $\Phi(x, y; \rho)$ be the corresponding distribution function. Using (5.73), show that

$$\Phi(x, y; \rho) = \Phi(x)\Phi(y) + \int_0^\rho \phi(x, y; \zeta)d\zeta. \tag{5.99}$$

[Section 5.7; Cramér and Leadbetter, 1967, pp. 26–7]

5.10 Slepian's inequality. Let $X_0 \sim \mathrm{MN}_m(0, [\kappa_0^{i,j}])$ and $X \sim \mathrm{MN}_m(0, [\kappa^{i,j}])$ and suppose that

$$\begin{aligned} \kappa_0^{i,i} &= \kappa^{i,i} \qquad (1 \leqslant i \leqslant m), \\ \kappa_0^{i,j} &\leqslant \kappa^{i,j} \qquad (1 \leqslant i < j \leqslant m). \end{aligned}$$

Using (5.73) show that for any $c \in R^m$

$$P(X_0 \leqslant c) \leqslant P(X \leqslant c).$$

[Section 5.7; Slepian, 1962]

5.11 Inclusion–exclusion formula and Möbius inversion. Let E be a finite set and let $\mathscr{P} = \mathscr{S}(E)$, the collection of all subsets of E, partially ordered by ordinary set inclusion. Show that

$$\mu(e, e') = (-1)^{|e'| - |e|}$$

for arbitrary subsets e and e' of E and with $|e|$ denoting the number of elements in e. Use this result together with Möbius inversion to prove the inclusion–exclusion formula of probability theory which states that

$$\begin{aligned} P\left(\bigcup_{i=1}^m A_i\right) = \sum_{i=1}^m P(A_i) - \sum_{1 \leqslant i < j \leqslant m} P(A_i \cap A_j) + \cdots \\ + (-1)^{m-1} P\left(\bigcap_{i=1}^m A_i\right). \end{aligned}$$

[Section 5.8]

5.12 The number of different partitions with k blocks of a set E with n elements is denoted by $S(n, k)$ and the quantities $S(n, k)$ are called *Stirling numbers of the second kind.* Show that they satisfy the basic recurrence relation

$$S(n, k) = kS(n-1, k) + S(n-1, k-1).$$

For some further information on these numbers, see Stanley (1986).

[Section 5.8]

5.13 Relation (5.83) exhibits a partition of the integer N into v parts. The number of such partitions is denoted by $p_v(N)$. Show that

$$p_v(N) = p_{v-1}(N-1) + p_v(N-v).$$

For further information, see Hall (1986).

[Section 5.9]

5.14 *The Hammersley–Clifford theorem.* For a very clear review of the key Hammersley–Clifford theorem on Markov random fields and potential functions, see Isham (1981). A simple, but typical and important, instance of that theorem is as follows.

Let $G = (\Gamma, E)$ be a finite connected graph with vertex set Γ and edge set E. For each $\gamma \in \Gamma$ let I_γ be a finite index set (the set of *levels* at γ) and let $i = \{i_\gamma \in I_\gamma : \gamma \in \Gamma\}$ denote a point in the product set $I = \prod_\gamma I_\gamma$. A probability (measure) p on I, with $p(i) > 0$ for all $i \in I$, is said to be *Markov* with respect to G if

$$i_A \text{ is independent of } i_{\Gamma\backslash A} \text{ given } i_{\partial A} \text{ for all } A \subset \Gamma. \tag{5.100}$$

Here, thinking of i as a random variable, we use i_A to denote the random variable $\{i_\gamma \in I_\gamma : \gamma \in A\}$, and (5.100) states that, under the probability distribution p, i_A and $i_{\Gamma\backslash A}$ are conditionally independent given the values on the boundary ∂A of A. The boundary of A is defined, in terms of the graph G, as the set of vertices $\gamma \in \Gamma\backslash A$ such that γ is connected to a point α in A by an edge in E, i.e. γ and α are *neighbours*.

Theorem (Hammersley–Clifford)

A probability p on I is Markov with respect to $G = (\Gamma, E)$ if and only if there exists functions ϕ_A, $A \subset \Gamma$, defined on I such that $\phi_A \equiv 0$ unless A is a complete subset of Γ and such that

$$\log p(i) = \sum\nolimits^{c} \phi_A(i). \tag{5.101}$$

In (5.101), $\sum^c$ indicates summation over all complete subsets of Γ, a complete subset (also called a clique) A being a collection of vertices all pairs of which are neighbours. The function $\phi_\cdot(\cdot)$ is termed a *nearest neighbour potential.*

This theorem, once considered very difficult to prove, can now be relatively easily established by means of Möbius inversion for the inclusion lattice, discussed in Exercise 5.11 (Lauritzen, 1982).

Statistical applications of versions of the Hammersley–Clifford theorem may be found as indicated:

models with discrete variates (contingency tables)	Darroch, Lauritzen and Speed (1980); Lauritzen (1982)
models with continuous variates	Kiiveri, Speed and Carlin (1984); Barndorff-Nielsen and Blæsild (1988a)
models with discrete and continuous variates	Lauritzen and Wermuth (1989)
spatial statistics	Isham (1981); Baddeley and Møller (1987)

[Section 5.8]

5.15 Factorial dispersion models. Hasse diagrams and Möbius inversion are very useful also in connexion with a general discussion of balanced factorial dispersion models, see Tjur (1984), Speed (1987).
[Section 5.8]

Bibliographic notes

Index notation is widely used in differential geometry and theoretical physics. The recognition of its substantial potential for use in statistics is due to McCullagh (1984, 1987). The fact that the relations between moments and cumulants constitute a special case of what we have termed the exlog relations was indicated and exemplified in McCullagh (1987). The exlog relations also occur more or less explicitly in statistical mechanics; see Ruelle (1969, section 4.4) and also Malyshev (1980). Speed (1983) introduced the technique of Möbius inversion and partition lattices in the study of cumulants. The important relation (5.51) can be traced to papers by James (1958), Leonov and Shiryaev (1959) and James and Mayne (1962).

Cumulants were first introduced and studied by the astronomer and statistician T.N. Thiele (1889, 1897, 1899) who called them *halvinvarianter* (i.e. semi-invariants); see Hald (1981) and Lauritzen (1981) for reviews of Thiele's work. Fisher (1929) rediscovered cumulants and developed their theory and applicability substantially.

The tensorial Hermite polynomials were introduced by Grad (1949). The exposition given here draws on Amari and Kumon (1983). The general result (5.71)–(5.72), which extends the well-known (5.73), appears to be new.

CHAPTER 6

Multivariate asymptotic expansions

6.1 Introduction

In Chapters 3 and 4 we outlined some of the main ideas connected with asymptotic expansions of one-dimensional integrals and series and of univariate distributions. Broadly the same ideas apply in the multidimensional case. Although some complications of notation are unavoidable we have aimed to show methods and results as direct generalizations of the corresponding univariate discussion.

One key method that extends directly to the multidimensional case is the Laplace expansion of integrals and we begin with a review of this together with some examples.

6.2 Laplace's method

In section 3.3 we gave in the univariate case an important method for the asymptotic expansion of integrals applying when the major contribution to the integral comes from a sharp peak in the integrand. The key to the argument was to produce in the integrand a factor close to a normal density of small variance. Essentially the same argument applies in m dimensions.

Consider as $n \to \infty$

$$w(n) = \int_D e^{-nr(y)} q(y) dy, \tag{6.1}$$

where the functions $r(y)$ and $q(y)$ are defined over the region D of integration, $D \subset R^m$.

As noted in section 3.3, the same arguments apply to more general integrals of the form

$$\int_D e^{-r_n(y)} q(y) dy \tag{6.2}$$

and indeed if, further, the factor $q(y)$ also depends on n but has, say, an asymptotic expansion in powers of $1/\sqrt{n}$, we may be able to argue term by term.

Suppose that in (6.1) $r(y)$ has a unique minimum in the interior of D at $y = \tilde{y}$ and that by taking second-degree terms of the Taylor expansion of $r(y)$ about $y = \tilde{y}$ we can write

$$e^{-nr(y)} = e^{-n\tilde{r}}\{\phi_m(y - \tilde{y}; \tilde{\kappa}/n)/\phi_m(0; \tilde{\kappa}/n)\}\{1 + O(1/\sqrt{n})\}, \quad (6.3)$$

where $\tilde{r} = r(\tilde{y})$ and $\phi_m(x; \kappa)$ is the density of the m-variate multivariate normal distribution of zero mean and covariance matrix κ, so that in particular

$$\phi_m(0; \kappa) = (2\pi)^{-(1/2)m}\{\det(\kappa)\}^{-1/2}.$$

The form of the Taylor expansion shows that

$$\tilde{\kappa}^{-1} = \{\nabla\nabla^{\mathrm{T}} r(y)\}_{y=\tilde{y}} = \tilde{r}'', \quad (6.4)$$

say. Thus $\det(\tilde{\kappa}) = 1/\det(\tilde{r}'')$. It follows on inserting (6.3) into (6.1) that the leading term of the expansion of (6.1) is

$$w(n) \sim \frac{e^{-n\tilde{r}}\tilde{q}(2\pi)^{(1/2)m}}{n^{m/2}\sqrt{\det(\tilde{r}'')}}, \quad (6.5)$$

where $\tilde{q} = q(\tilde{y})$, in direct generalization of (3.16). The relative error in this approximation is of order $O(n^{-1})$. Detailed calculation shows that the term in $n^{-1/2}$ vanishes, essentially because the normal distribution being exploited has zero mean.

The regularity conditions for this are given in section 6.12.

In the above discussion it is assumed that the minimum of $r(y)$, and hence the major contribution to the integral, is interior to the region of integration. As a simple illustration of the modification necessary if the minimum is on the boundary, consider

$$\int_{-\infty}^{\infty} dx \int_{0}^{\infty} dy e^{-nr(x,y)} q(x, y) \quad (6.6)$$

and suppose for simplicity that for each x, $r(x, y)$ has its minimum at $y = 0$ with nonzero slope there, $\partial r(x, 0)/\partial y > 0$. Then the integral with respect to y is, by the argument of section 3.3,

$$e^{-nr(x,0)}\{n\partial r(x, 0)/\partial y\}^{-1}\{1 + O(n^{-1})\}. \quad (6.7)$$

Now suppose that $r(x, 0)$ has its minimum at $x = \tilde{x}$ with nonzero second derivative there, $\partial^2 r(\tilde{x},0)/\partial x^2 > 0$. Then by a second expansion, this time with respect to x, (6.6) has for large n the form

$$\frac{\sqrt{(2\pi)}e^{-nr(\tilde{x},0)}q(\tilde{x},0)}{n^{3/2}\{\partial r(\tilde{x},0)/\partial y\}\{\partial^2 r(\tilde{x},0)/\partial x^2\}^{1/2}}. \tag{6.8}$$

It is clear that as regards the behaviour with respect to y only that locally near $x = \tilde{x}$ is relevant. The argument extends immediately if x is multidimensional.

Higher-order terms in the expansion of (6.1) or (6.2) are easily determined by means of the exlog relations.

Suppose, for instance, that $r_n(y)$ of (6.2) has a uniquely determined minimum $\tilde{y}$ in the interior of D. We allow this minimum to depend on n, but $\tilde{y}$ should then converge sufficiently rapidly to an interior point of D as $n \to \infty$. Making a Taylor expansion of $r_n(y)$ around $\tilde{y}$, we have

$$r_n(y) = \tilde{r}_n + \sum_{\nu=2}^{\infty} \frac{1}{\nu!}(y-\tilde{y})_{i_1\ldots i_\nu}\tilde{r}_n^{(i_1\ldots i_\nu)},$$

where $(y-\tilde{y})_{i_1\ldots i_\nu} = (y-\tilde{y})_{i_1}\ldots(y-\tilde{y})_{i_\nu}$. Hence, by the exlog relation (5.27),

$$e^{-r_n(y)} = (2\pi)^{(1/2)m}|\det(\tilde{\kappa}_n)|^{1/2}e^{-\tilde{r}_n}\phi(y-\tilde{y};\tilde{\kappa}_n)\Big\{1 - \frac{1}{3!}(y-\tilde{y})_{ijk}\tilde{r}_n^{(ijk)}$$
$$- \frac{1}{4!}(y-\tilde{y})_{ijkl}\tilde{r}_n^{(ijkl)} - \frac{1}{5!}(y-\tilde{y})_{ijklp}\tilde{r}_n^{(ijklp)}$$
$$- \frac{1}{6!}(y-\tilde{y})_{ijklpq}(\tilde{r}_n^{(ijklpq)} - 10\tilde{r}_n^{(ijk)}\tilde{r}_n^{(lpq)}) - \cdots\Big\},$$

where $\tilde{\kappa}_n = [\tilde{\kappa}_{nij}]$ is the inverse matrix of $[\tilde{r}_n^{(ij)}]$. Letting, for simplicity, $q(y) \equiv 1$, we find that, in wide generality, the first terms of a valid asymptotic expansion of the integral of $\exp\{-r_n(y)\}$ are given by

$$\int_D e^{-r_n(y)}dy \sim (2\pi)^{(1/2)m}|\det(\tilde{r}_n'')|^{-1/2}e^{-\tilde{r}_n}(1 + C_{1n}),$$

where

$$C_{1n} = -\tfrac{1}{24}\bar{R}_{4n} + \tfrac{1}{72}(9\bar{R}_{13n} + 6\bar{R}_{23n}) \tag{6.9}$$

with

$$\bar{R}_{13n} = \tilde{r}_n^{(ijk)}\tilde{r}_n^{(lpq)}\tilde{\kappa}_{nij}\tilde{\kappa}_{nkl}\tilde{\kappa}_{npq}, \tag{6.10}$$

$$\bar{R}_{23n} = \tilde{r}_n^{(ikp)}\tilde{r}_n^{(jlq)}\tilde{\kappa}_{nij}\tilde{\kappa}_{nkl}\tilde{\kappa}_{npq}, \tag{6.11}$$

$$\bar{R}_{4n} = \tilde{r}_n^{(ijkl)}\tilde{\kappa}_{nijkl}. \tag{6.12}$$

When $r_n(y)$ is of the form $nr(y)$ the asymptotic expansion of the integral of $\exp\{-r_n(y)\}$ is in powers of n^{-1}. In particular, the correction term in (6.9) is then of the form $C_{1n} = n^{-1}C_1$ with

$$C_1 = -\tfrac{1}{24}R_4 + \tfrac{1}{72}(9R_{13} + 6R_{23}), \tag{6.13}$$

where

$$R_{13} = \tilde{r}^{(ijk)}\tilde{r}^{(lpq)}\tilde{\kappa}_{ij}\tilde{\kappa}_{kl}\tilde{\kappa}_{pq}, \tag{6.14}$$

$$R_{23} = \tilde{r}^{(ikp)}\tilde{r}^{(jlq)}\tilde{\kappa}_{ij}\tilde{\kappa}_{kl}\tilde{\kappa}_{pq}, \tag{6.15}$$

$$R_4 = \tilde{r}^{(ijkl)}\tilde{\kappa}_{ijkl}, \tag{6.16}$$

in the obvious notation.

Example 6.1 A double integral related to the beta function

As a simple example that illustrates some general issues, consider the function

$$I_n(a_1, a_2, b; s_1, s_2) = \int_0^1 dy_1 \int_0^1 dy_2 y_1^{na_1} y_2^{na_2}(1 - \tfrac{1}{2}y_1 - \tfrac{1}{2}y_2)^{nb} e^{-s_1y_1 - s_2y_2}. \tag{6.17}$$

Note that the integrand has the same sign at all points in the region of integration; that this should be essentially true is a requirement for Laplace's method to be sensible.

Now for large n we can apply Laplace's method in several ways of which the most natural are:

1. to use the special form (6.1) directly, thus taking $q(y) = e^{-s_1y_1 - s_2y_2}$;
2. to absorb $q(y)$ into the function $r_n(y)$ by writing $r_n(y) = -na_1 \log y_1 - na_2 \log y_2 - nb \log(1 - \tfrac{1}{2}y_1 - \tfrac{1}{2}y_2) + s_1y_1 + s_2y_2$.

If we use (2), the minimum occurs at $(\tilde{y}_1, \tilde{y}_2)$ where

$$(a_1 + b)\tilde{y}_1 + a_1\tilde{y}_2 = 2a_1 - n^{-1}\tilde{y}_1(2 - \tilde{y}_1 - \tilde{y}_2)s_1,$$
$$a_2\tilde{y}_1 + (a_2 + b)\tilde{y}_2 = 2a_2 - n^{-1}\tilde{y}_2(2 - \tilde{y}_1 - \tilde{y}_2)s_2.$$

These equations are conveniently solved iteratively by first omitting the term in n^{-1} thereby obtaining the minimum under (1). Of course if $a_1 = a_2$, $s_1 = s_2$ then $\tilde{y}_1 = \tilde{y}_2$ and the equations simplify.

Evaluation of the second and indeed higher derivatives is straightforward and in fact

$$\begin{aligned}\tilde{r}_n^{(11)} &= na_1/\tilde{y}_1^2 + \tfrac{1}{4}nb/(1 - \tfrac{1}{2}\tilde{y}_1 - \tfrac{1}{2}\tilde{y}_2)^2,\\ \tilde{r}_n^{(12)} &= \tilde{r}_n^{(21)} = \tfrac{1}{4}nb/(1 - \tfrac{1}{2}y_1 - \tfrac{1}{2}y_2)^2,\\ \tilde{r}_n^{(22)} &= na_2/\tilde{y}_2^2 + \tfrac{1}{4}nb/(1 - \tfrac{1}{2}\tilde{y}_1 - \tfrac{1}{2}\tilde{y}_2)^2.\end{aligned}$$

6.3 Edgeworth expansions

In section 4.2 we showed how the Edgeworth series arises naturally via series expansion of the moment generating function of standardized sums of independent random variables. We then saw in section 4.5 that the expansion arises more generally. We follow the same broad path in the multidimensional case, using the tensorial notation introduced in section 5.3.

Let $Y_1,\ldots,Y_n$ be independent and identically distributed m-dimensional random variables, copies of a random variable $Y = (Y^1,\ldots,Y^m)$ with mean $\mu = (\mu^1,\ldots,\mu^m)$ and cumulants $\kappa^{i_1,\ldots,i_\nu} = K\{Y^{i_1},\ldots,Y^{i_\nu}\}$. Let

$$S_n = Y_1 + \cdots + Y_n, \qquad S_n^* = (S_n - n\mu)/\sqrt{n}. \tag{6.18}$$

Note that for a general treatment we do not carry out the further standardization to achieve unit variances and zero covariances. In dealing with the general formulation a further transformation would complicate rather than simplify the resulting formulae, in effect by leading to the omission of particular terms from expansions initially of very symmetrical structure. On the other hand, when it comes to detailed calculation with very specific cases, it will be wise to exploit any special features available and, in particular, any simple transformation to zero covariance should be made. See also the discussion of conditional distributions in section 7.1, where one random variable is replaced by its residual from linear least squares regression on another random variable.

The cumulant generating function of S_n^* is

$$\begin{aligned}K(S_n^*; t) &= nK(Y - \mu; t/\sqrt{n})\\ &= \sum_{\nu=2}^{\infty} \frac{1}{\nu!}\kappa^{i_1,\ldots,i_\nu} t_{i_1}\ldots t_{i_\nu} n^{-\nu/2+1}.\end{aligned} \tag{6.19}$$

In (6.19) the leading term with $\nu = 2$ has order 1 and corresponds to the asymptotic multivariate normal density of S_n^* of zero mean and covariance matrix $\kappa = (\kappa^{i,j})$, namely $\phi_m(x;\kappa)$.

Thus in the two-dimensional case the terms for $\nu = 2, 3$ in (6.19) give

$$\tfrac{1}{2}(\kappa^{1,1}t_1^2 + 2\kappa^{1,2}t_1t_2 + \kappa^{2,2}t_2^2) + \frac{1}{6\sqrt{n}}(\kappa^{1,1,1}t_1^3 + 3\kappa^{1,1,2}t_1^2t_2 + 3\kappa^{1,2,2}t_1t_2^2 + \kappa^{2,2,2}t_2^3), \quad (6.20)$$

where, for example, $\kappa^{1,2} = E\{(Y^1 - \mu^1)(Y^2 - \mu^2)\} = \operatorname{cov}(Y^1, Y^2)$, $\kappa^{1,1,1} = \mu_3(Y^1) = \kappa_3(Y^1)$, etc.

We now obtain an asymptotic expansion for the density of S_n^*, in powers of $1/\sqrt{n}$ by taking the first few terms of (6.19), converting to an expansion for the moment generating function $M(S_n^*;t)$ and then inverting term by term. Thus taking terms to order $1/n$, the most that we commonly require in statistical applications,

$$\begin{aligned} M(S_n^*;t) &= \exp\{K(S_n^*;t)\} \\ &= \exp(\tfrac{1}{2}\kappa^{i,j}t_it_j)\left\{1 + \frac{1}{6\sqrt{n}}\kappa^{i,j,k}t_it_jt_k + \frac{1}{24n}\kappa^{i,j,k,l}t_it_jt_kt_l \right. \\ &\quad \left. + \frac{1}{72n}\kappa^{i,j,k}\kappa^{l,p,q}t_it_jt_kt_lt_pt_q\right\} + O(n^{-3/2}). \end{aligned} \quad (6.21)$$

We thus need the inverse of the product of the exponential term in (6.21) with the polynomial factors and these are given in terms of the tensorial Hermite polynomials; cf. section 5.7. With the aid of the covariant Hermite polynomials we write

$$\begin{aligned} f(S_n^*;x) &= \phi_m(x;\kappa)\left\{1 + \frac{1}{6\sqrt{n}}\kappa^{i,j,k}h_{ijk}(x;\kappa) + \frac{1}{24n}\kappa^{i,j,k,l}h_{ijkl}(x;\kappa) \right. \\ &\quad \left. + \frac{1}{72n}\kappa^{i,j,k}\kappa^{l,p,q}h_{ijklpq}(x;\kappa)\right\} + O(n^{-3/2}), \end{aligned} \quad (6.22)$$

and the formal parallel with the univariate form (4.6) is clear.

The general form of the Edgeworth expansion is

$$\begin{aligned} &\phi_m(x;\kappa)\{1 + Q_3(x;\kappa^{(3)})/\sqrt{n} + Q_4(x;\kappa^{(4)})/n + \cdots + Q_\nu(x;\kappa^{(\nu)})/n^{(1/2)\nu - 1}\} \\ &\quad + O(n^{-(1/2)\nu + (1/2)}), \end{aligned} \quad (6.23)$$

where $Q_\nu(x;\kappa^{(\nu)})$ is a polynomial of degree ν containing cumulants up to order ν.

The simplest nontrivial example of such an expansion takes $\nu = 3$,

$n=2$. For this we take (6.22), writing

$$f(S_n^*; x) = \phi_2(x;\kappa)\left\{1 + \frac{1}{6\sqrt{n}}\kappa^{i,j,k}h_{ijk}(x;\kappa)\right\} + O(n^{-1}). \quad (6.24)$$

The 8 terms in the sum $\kappa^{i,j,k}h_{ijk}(x;\kappa)$ are

$$\begin{aligned}&\kappa^{1,1,1}h_{111}(x;\kappa) + 3\kappa^{1,1,2}h_{112}(x;\kappa) + 3\kappa^{1,2,2}h_{122}(x;\kappa)\\&\quad + \kappa^{2,2,2}h_{222}(x;\kappa).\end{aligned} \quad (6.25)$$

Further,

$$\begin{aligned}h_{111}(x;\kappa) &= h^{ijk}(x;\kappa)\kappa_{i,1}\kappa_{j,1}\kappa_{k,1}\\&= h^{111}(x;\kappa)\kappa_{1,1}^3 + 3h^{112}(x;\kappa)\kappa_{1,1}^2\kappa_{2,1}\\&\quad + 3h^{122}(x;\kappa)\kappa_{1,1}\kappa_{2,1}^2 + h^{222}(x;\kappa)\kappa_{2,1}^3,\\h_{112}(x;\kappa) &= h^{ijk}(x;\kappa)\kappa_{i,1}\kappa_{j,1}\kappa_{k,2}\\&= h^{111}(x;\kappa)\kappa_{1,1}^2\kappa_{1,2} + h^{112}(x;\kappa)(2\kappa_{1,1}\kappa_{1,2}^2 + \kappa_{1,1}^2\kappa_{2,2})\\&\quad + h^{122}(x;\kappa)(2\kappa_{1,1}\kappa_{2,2}\kappa_{1,2} + \kappa_{1,2}^3) + h^{222}(x;\kappa)\kappa_{1,2}^2\kappa_{2,2},\end{aligned}$$

with analogous expressions for $h_{122}(x;\kappa)$ and $h_{222}(x;\kappa)$. Here the superscripts to the κ's are powers and not identifiers.

This serves to illustrate the relative complexity of detail subsumed in the general formula (6.23). As a special case suppose that the components of X are uncorrelated and of unit variance, so that $\kappa^{1,1}=\kappa_{1,1}=\kappa^{2,2}=\kappa_{2,2}=1$, $\kappa^{1,2}=\kappa_{1,2}=0$. Then

$$\begin{aligned}h_{111}(x;I) &= h^{111}(x;I) = H_3(x^1),\\h_{112}(x;I) &= h^{112}(x;I) = H_2(x^1)H_1(x^2),\\h_{122}(x;I) &= h^{122}(x;I) = H_1(x^1)H_2(x^2),\\h_{222}(x;I) &= h^{222}(\kappa;I) = H_3(x^2)\end{aligned} \quad (6.26)$$

and the correction term (6.25) reduces to

$$\begin{aligned}&\kappa^{1,1,1}H_3(x^1) + 3\kappa^{1,1,2}H_2(x^1)H_1(x^2)\\&\quad + 3\kappa^{1,2,2}H_1(x^1)H_2(x^2) + \kappa^{2,2,2}H_3(x^2).\end{aligned}$$

This allows a fairly direct interpretation of the third-order cumulants in this special case.

So far we have discussed Edgeworth expansions for sums of independent and identically distributed random vectors all of whose cumulants are known. The expansion holds much more generally, however; for instance, if the defining cumulants are not available exactly they can be replaced by asymptotic expansions with consequent modification of the detailed formulae; see further sections 6.7 and 6.10.

In view of this it is sometimes convenient to work with a slightly different notation. For any m-dimensional random variable X of mean 0 and cumulants $\kappa^{i,j,\dots}$, and for any $r=0,1,\dots$ let

$$p^{[r]}(x)=p^{[r]}(X;x)=\phi_m(x;\kappa)\{1+Q_3(x;\kappa)+\cdots+Q_{r+2}(x;\kappa)\}, \tag{6.27}$$

where $\kappa=[\kappa^{i,j}]$ is the covariance matrix of X and

$$\begin{aligned} Q_3(x;\kappa)&=\tfrac{1}{6}\kappa^{i,j,k}h_{ijk}(x;\kappa),\\ Q_4(x;\kappa)&=\tfrac{1}{24}\kappa^{i,j,k,l}h_{ijkl}(x;\kappa)+\tfrac{1}{24}\kappa^{i,j,k}\kappa^{l,m,n}h_{ijklmn}(x;\kappa), \end{aligned} \tag{6.28}$$

etc.

Then, due to the tensorial nature of cumulants and of the Hermite polynomials h, (6.22) may be re-expressed as

$$f(S_n^*;x)=p^{[r]}(S_n^*;x)+O(n^{-(1/2)r-(1/2)}). \tag{6.29}$$

In this way the dependence of (6.22) on n has been absorbed in the notation (6.27)–(6.29).

Irrespective of the context, (6.27) can be contemplated as an approximation to the distribution of the random vector X, but in any given case the nature of this approximation, whether in an asymptotic sense or otherwise, needs careful consideration. In most detailed calculations the inclusion of the factors $n^{-1/2}, n^{-1},\dots$ is an aid to the consistent retention of terms in making expansions and it is for that reason we have for most purposes preferred the more explicit notation.

Various theorems necessary for a rigorous account of Edgeworth expansions are discussed in section 6.13.

Example 6.2 Convolution of bivariate log normal distributions

Let $Y_1,\dots,Y_n$ be n independent and identically distributed bivariate random variables, copies of a random variable $Y=(Y^1,Y^2)$ having the bivariate log normal distribution of Example 5.6. Then we can write down an Edgeworth expansion for the distribution of

$$S_n^* = \frac{Y_1 + \cdots + Y_n - n(\kappa^1, \kappa^2)}{\sqrt{n}} = \frac{S_n - n(\kappa^1, \kappa^2)}{\sqrt{n}}$$

in terms of the standardized cumulants and the tensorial Hermite polynomials associated with the covariance matrix

$$\begin{pmatrix} \kappa^{1,1} & \kappa^{1,2} \\ \kappa^{2,1} & \kappa^{2,2} \end{pmatrix}$$

defined via (5.54). This gives as leading term in the expansion for S_n a normal distribution with the correct mean and covariance matrix. It is tempting to try and amalgamate the $1/\sqrt{n}$ term of the Edgeworth series into the leading term, by taking as the leading term a log normal distribution with the correct mean and covariance matrix. This could be explored in detail numerically but is most simply studied via the standardized third cumulants. Thus for the marginal standardized third cumulant of the first component the ratio of the value for S_n and the value for the approximate log normal distribution is, after some calculation,

$$\frac{e^{\sigma_{20}} + 2}{3 + n^{-1}(e^{\sigma_{20}} - 1)} \tag{6.30}$$

and for the corresponding $\kappa^{1,1,2}$ the standardized ratio is, for $\sigma_{11} \neq 0$,

$$\frac{e^{\sigma_{20}+\sigma_{11}} + e^{\sigma_{20}} - 2}{e^{\sigma_{11}} + 2e^{\sigma_{20}} - 3 + n^{-1}(e^{\sigma_{20}} - 1)(e^{\sigma_{11}} - 1)}. \tag{6.31}$$

The first ratio exceeds 1 and the second does so if $\sigma_{11} > 0$. The easiest way to obtain some quantitative idea of their values is to assume that the σ_{ij} are reasonably small, to expand as far as quadratic terms to give for (6.30) and (6.31) respectively for fixed n:

$$1 + \tfrac{1}{3}\sigma_{20}(1 - 1/n) + o(\sigma_{20}),$$

$$1 + \frac{\sigma_{20}\sigma_{11}(1 - 1/n)}{\sigma_{11} + 2\sigma_{20}} + o(\sigma).$$

This is another instance of so-called small dispersion asymptotics.

It follows that use of the log normal distribution as the leading term is a substantial improvement on the normal distribution but that it still underestimates both the marginal skewness and, unless

σ_{02} is relatively large, those properties encapsulated in the mixed third cumulants.

6.4 Exponential models

The exponential family of models, or more briefly exponential models, play a central role in parametric statistical theory and, as we have seen in section 4.3, are also a valuable mathematical tool in deriving asymptotic expansions. It is convenient to outline here some of the mathematical properties of such models; for a full treatment, see Barndorff-Nielsen (1978, 1988).

Suppose that we have a parameter ω taking values in Ω and that $\theta(\omega)$ and $s(x)$ are vectors of common dimension, m. We consider the family, $\mathcal{M}$, of densities with respect to a suitable measure μ, usually counting measure or Lebesgue measure:

$$p(x;\theta)d\mu = \exp\{\theta(\omega)\cdot s(x) - k\{\theta(\omega)\} + a(x)\}d\mu. \qquad (6.32)$$

Then $\theta(\omega)$ is called the *canonical parameter* and $s = s(x)$ the *canonical statistic*. Generic coordinates for θ and s will be denoted by θ^i and s_i, so that $\theta\cdot s = \theta^i s_i$.

Among the many possible representations of a given family of models, we choose one with the smallest value of m; any such representation is said to be *minimal* and under such a representation s is minimal sufficient. Here we take (6.32) to be minimal throughout.

Now for $\omega\in\Omega$, θ lies in some space Θ. Associated with (6.32) also is, however, the model

$$\exp\{\theta\cdot s - k(\theta) + a(x)\}d\mu,$$

where θ is constrained only by the requirement of convergence

$$\int \exp\{\theta\cdot s(x) + a(x)\}\,d\mu < \infty.$$

The resulting set of θ values, denoted by $\tilde{\Theta}$, is a convex subset of R^m, whereas in general Θ is a subset of $\tilde{\Theta}$ determined by Ω.

We call the model $\tilde{\mathcal{M}}$

$$p(x;\theta)d\mu = \exp\{\theta\cdot s - k(\theta) + a(x)\}\,d\mu \qquad (\theta\in\tilde{\Theta}), \qquad (6.33)$$

the *full exponential family* generated by $\mathcal{M}$. In virtue of the normalization, we have that

$$k(\theta) = \log \int \exp\{\theta \cdot s + a(x)\} d\mu. \tag{6.34}$$

Thus $\tilde{\mathcal{M}}$ is an exponential model containing $\mathcal{M}$ as a submodel. In cases of most common concern Θ is a d-dimensional submanifold of $\tilde{\Theta}$; then $\mathcal{M}$ is called an (m, d) exponential family.

For example, if $X = (X_1, X_2)$ is such that X_1 and X_2 are independently normally distributed with unit variance and with, in $\mathcal{M}$, $E(X_1) = \omega$, $E(X_2) = c\omega^2$, for a known constant $c \neq 0$, then $\mathcal{M}$ is a $(2, 1)$ family. In $\tilde{\mathcal{M}}$, the means of X_1 and X_2 are arbitrary. Thus $\tilde{\Theta}$ is R^2, whereas Θ is a parabola in R^2.

Now it is easily shown that under any $\theta \in \Theta$, the cumulant generating function of s is

$$\log E(e^{tS}; \theta) = k(\theta + t) - k(\theta). \tag{6.35}$$

It is useful to call $k(\theta)$ the *cumulant transform* of $\mathcal{M}$ or of the random variable S. In fact,

$$k_{i_1 \ldots i_v}(\theta) = \frac{\partial^v k(\theta)}{\partial \theta^{i_1} \cdots \partial \theta^{i_v}} \tag{6.36}$$

is the elemental cumulant of $S_{i_1}, \ldots, S_{i_v}$ evaluated under the distribution specified by θ, i.e.

$$k_{i_1 \ldots i_v}(\theta) = K_\theta\{S_{i_1}, \ldots, S_{i_v}\}. \tag{6.37}$$

It is of special importance to consider the mean value of S as a function of θ for $\theta \in \operatorname{ing}(\tilde{\Theta})$, the interior of $\tilde{\Theta}$. We write

$$\tau(\theta) = E(S; \theta) = \nabla k(\theta) \tag{6.38}$$

and call τ the *mean value* parameter of the family. The transformation from θ to $\tau(\theta)$ is a one-to-one correspondence, in fact a diffeomorphism, between $\operatorname{int}(\tilde{\Theta})$ and $\mathcal{S}$, say. The generic coordinates of τ are written τ_i. It can be shown that $\mathcal{S}$ is contained in the interior of the closed convex hull, C, of the support of S.

The model $\tilde{\mathcal{M}}$ is called *regular* if $\tilde{\Theta}$ is open and called *steep* if $|\tau(\theta)| \to \infty$ for any sequence of points $\theta \in \operatorname{int}(\tilde{\Theta})$ tending to a boundary point of Θ along a line segment; a regular model is automatically steep (Barndorff-Nielsen, 1978). As an example of a nonsteep model, consider the family of probability density functions on the interval

$1 \leqslant x < \infty$ given by

$$e^{\theta x - k(\theta) - c \log x},$$

where $c > 2$ is viewed as a known constant and the variation domain of the canonical parameter θ is $(-\infty, 0]$. An important example of a steep but nonregular exponential model is provided by the family of inverse Gaussian distributions.

We call $\mathscr{M}$ a *core* exponential model if $\tilde{\mathscr{M}}$ is steep and $\Theta = \text{int}(\tilde{\Theta})$. The following theorem summarizes many key properties of this important class of models. In this theorem we refer to Legendre transforms which are reviewed briefly in section 6.14.

Theorem 6.1

If $\mathscr{M}$ is a core exponential model, then:

(i) The space $\mathscr{S}$ of possible values of the mean value parameter is the same as int C, the interior of the closed convex hull of the support of the canonical statistic S.

(ii) The cumulant transform k is a strictly convex function on Θ. Its Legendre transform $k^*(s)$ also is strictly convex, this time on $\mathscr{S}$, and can be interpreted as the maximized log likelihood function $\hat{l}(s)$

$$k^*(s) = \hat{l}(s) \qquad (s \in \mathscr{S}), \tag{6.39}$$

where

$$\hat{l}(s) = \sup_{\theta \in \Theta} \{\theta \cdot s - k(\theta)\}. \tag{6.40}$$

(iii) Further, the Legendre transform of $k^*(s)$ satisfies

$$k^{**}(\theta) = \sup_{s \in \mathscr{S}} \{\theta \cdot s - k^*(s)\} = k(\theta). \tag{6.41}$$

(iv) Both k and k^* can be differentiated infinitely often and, writing $\Sigma = \text{cov}(S; \theta)$, we have

$$\begin{aligned} \nabla_\theta k &= \tau, & \nabla_\tau k^* &= \theta, \\ \nabla_\theta \nabla_\theta^{\mathrm{T}} k &= \Sigma, & \nabla_\tau \nabla_\tau^{\mathrm{T}} k^* &= \Sigma^{-1}. \end{aligned} \tag{6.42}$$

(v) The maximum likelihood estimate $\hat{\theta}$ exists if and only if

$s \in \mathscr{S} = \operatorname{int} C$ and then is the unique solution of the maximum likelihood estimating equation

$$E(S; \theta) = s. \tag{6.43}$$

Results (6.42) and (6.43) are of immediate wide applicability.

6.5 Tilted expansions

We introduce the multivariate tilted approximation as an indirect Edgeworth expansion, as in section 4.3.

As in section 6.3 the discussion is strongly motivated by the behaviour of sums of independent random variables. To emphasize the greater generality we shall, however, denote the random variable under study by X_n, rather than S_n, with density $f_n(X_n; x) = f_n(x)$, say. We wish to approximate to $f_n(x)$ at an assigned point x. We introduce the exponential family generated by $f_n(x)$ in the form

$$f_n(x; \lambda) = \exp\{\lambda \cdot x - k_n(\lambda)\} f_n(x), \tag{6.44}$$

where the normalizing constant $k_n(\lambda)$ is the cumulant generating function of $f_n(x)$. This is the exponential tilt of the distribution $f_n(x)$. Now, for an arbitrary $\tilde{\lambda}$,

$$\begin{aligned} f_n(x) &= f_n(x; \tilde{\lambda}) \frac{f_n(x)}{f_n(x; \tilde{\lambda})} \\ &= f_n(x; \tilde{\lambda}) \exp\{k_n(\tilde{\lambda}) - \tilde{\lambda} \cdot x\} \end{aligned} \tag{6.45}$$

and the key idea of exponential tilting is that we choose $\tilde{\lambda}$ so that some good approximation is available for $f_n(x; \tilde{\lambda})$.

If that approximation is an Edgeworth series it will be sensible to choose $\tilde{\lambda}$ so that to a close approximation $E(X_n; \tilde{\lambda}) = x$ and this requires that $\tilde{\lambda}$ is close to $\hat{\lambda}_n(x)$, the formal maximum likelihood estimate of λ in the statistical model (6.44), when x is observed. This is because the cumulant generating function of (6.44) for given λ is

$$k_n(\lambda + t) - k_n(\lambda), \tag{6.46}$$

so that $E(X_n; \lambda) = \nabla k_n(\lambda)$ and hence $\hat{\lambda}_n(x)$ satisfies the vector equation

$$[\nabla k_n(\lambda)]_{\lambda = \hat{\lambda}_n(x)} = x. \tag{6.47}$$

Under suitable regularity conditions, the leading term of the resulting Edgeworth expansion is a multivariate normal density evaluated at its mean and is thus found in terms of the determinant of the covariance matrix. By (6.46) this determinant is

$$\det\{\nabla\nabla^{\mathrm{T}}k_n(\lambda)\}_{\lambda=\hat{\lambda}_n(x)} = \det\{j_n(\hat{\lambda})\}, \tag{6.48}$$

where $j_n(\hat{\lambda})$ denotes the formal observed and expected information about λ in the statistical model (6.44), evaluated at $\lambda = \hat{\lambda}$. Thus finally, from (6.45), the leading term of the expansion for $f_n(x)$ is

$$\frac{\exp\{k_n(\hat{\lambda}) - \hat{\lambda}\cdot x\}}{(2\pi)^{(1/2)m}[\det\{j_n(\hat{\lambda})\}]^{1/2}}. \tag{6.49}$$

This is called the *unnormalized tilted approximation.*

Because we are using an Edgeworth expansion at its mean, the error in (6.49) is $O(n^{-1})$ and is a relative error for x in a large deviation region rather than an absolute error. If we include higher terms in the Edgeworth expansion for $f_n(x; \hat{\lambda})$, we obtain an expansion in powers of $1/n$, rather than in powers of $1/\sqrt{n}$ for the ordinary Edgeworth series, the error in stopping at any stage being a relative one. In an extension of the notation of (6.23) the general form is

$$\frac{\exp\{k_n(\hat{\lambda}) - \hat{\lambda}\cdot x\}}{(2\pi)^{(1/2)m}[\det\{j_n(\hat{\lambda})\}]^{1/2}}\left\{1 + \frac{Q_4(0; \hat{\kappa}^{(4)})}{n} + \frac{Q_6(0; \hat{\kappa}^{(6)})}{n^2} + \cdots\right\}, \tag{6.50}$$

where the Q's are the polynomials of section 6.3 evaluated at the origin and with coefficients determined under the distribution (6.44) at $\lambda = \hat{\lambda}$.

The leading term (6.49) is in general not exactly normalized, i.e. does not integrate to one over the support of x (or over R^m). It is therefore tempting to replace (6.49) by

$$c_n \frac{\exp\{k_n(\hat{\lambda}) - \hat{\lambda}\cdot x\}}{(2\pi)^{(1/2)m}[\det\{j_n(\hat{\lambda})\}]^{1/2}} \tag{6.51}$$

and to determine c_n, for example numerically, to achieve exact normalization; the modified form (6.51) is called the *normalized tilted approximation.* If the term $Q_4(0; \hat{\kappa}^{(4)})$ in (6.50) were independent of x the normalization would absorb that factor and the relative error of (6.51) would be $O(n^{-2})$. Usually, however, $Q_4(0; \kappa^{(4)})$ will depend on x; roughly speaking, if it does so gently over the range of x of interest, the normalized approximation will be an improvement, but if the term fluctuates strongly there is clearly the possibility that the normalized form performs poorly over part of the range.

The theorems that justify the results (6.49) and (6.50) rigorously for sums of independent and identically distributed random vectors are set out in section 6.13.

Let us consider the question of when an exponential model

$$e^{\theta\cdot x - k(\theta) + a(x)} \tag{6.52}$$

is such that (6.52) is exactly equal to its own normalized tilted approximation for every fixed θ. We assume here that (6.52) is a core exponential model, as defined in section 6.4, that the domain of variation of x equals the open convex set $\mathscr{T}$ of possible mean values of x under the model (6.52), and that (6.52) is the probability density function of x with respect to Lebesgue measure on $\mathscr{T}$. The normalized tilted approximation to the density (6.52) of x is of the form

$$c(\theta)[\det\{\nabla\nabla^{\mathrm{T}}k(\hat{\theta})\}]^{-1/2}e^{\theta\cdot x - k(\theta) - \hat{\theta}\cdot x + k(\hat{\theta})}. \tag{6.53}$$

For this to be exact for all θ it is necessary and sufficient that $\hat{\theta}\cdot x - k(\hat{\theta}) + \frac{1}{2}\log\det\{\nabla\nabla^{\mathrm{T}}k(\hat{\theta})\}$ is independent of x (Barndorff-Nielsen, 1988, section 7.1).

By Further results and exercises 4.15, if the order m of the exponential model (6.52) is 1 then the only cases of exactness are the normal distribution with known variance, the gamma distribution with known shape parameter, and the inverse Gaussian distribution with known precision. However, as the order m increases the exactness cases proliferate. In particular, under the multivariate normal model with both mean vector and covariance matrix unknown, the distribution of the minimal sufficient and canonical statistic has this property, and the same is true of many other of the most important models for multivariate normal analysis (regression, analysis of variance, etc.). Another class of examples derives from the inverse Gaussian distribution, illustrated in Example 6.3 below. For further discussion and examples of exactness, see Barndorff-Nielsen (1983, 1988), Barndorff-Nielsen and Blæsild (1988a), and also Further results and exercises 6.5.

Example 6.3 Inverse Gaussian distribution

Let $X_1, \ldots, X_n$ be a sample from the inverse Gaussian distribution, $N^-(\chi, \psi)$, with density

$$\frac{\sqrt{\chi}}{\sqrt{(2\pi)}}e^{\sqrt{(\chi\psi)}}x^{-3/2}e^{-1/2(\chi x^{-1} + \psi x)}. \tag{6.54}$$

For any $n>1$ and any $\chi>0$ and $\psi>0$ the normalized tilted approximation to the distribution of the minimal sufficient and canonical statistic $(\bar{X},\tilde{X})=n^{-1}(\sum X_i,\sum X_i^{-1})$ is exact, as follows by straightforward calculation, using the well-known facts that $\bar{X}$ and $\tilde{X}-\bar{X}^{-1}$ are independent, with $\bar{X}\sim N^{-}(n\chi,n\psi)$ and $n(\tilde{X}-\bar{X}^{-1})\sim \Gamma(\frac{1}{2}n-\frac{1}{2},\frac{1}{2}\chi)$.

6.6 Large deviations

Large deviation theory has as its initial object the asymptotic characterization of probabilities of the type $P(n^{-1}S_n\in B)$, where $S_n=X_1+\cdots+X_n$ is the sum of independent identically distributed random variates which are m-dimensional vectors or of some other, more abstract, nature and where B is a fixed, possibly infinitesimal, set. More generally, one studies the asymptotic behaviour of probabilities $P(n^{-1}T_n\in B)$, where T_n is a sequence of random variables whose limit properties are in some way similar to those of a sequence of sums S_n. In most of the literature on the subject the interest focuses on cases where the set B does not contain the centre of the distribution of T_n, this being typically expressed as $n^{-1}E(T_n)\notin B$ or as $T_n-E(T_n)$ being of order $O(n)$, whence the name of large deviations.

For the applications to statistical inference that we have in mind, however, it is essential to have results that give precise information also for $T_n-E(T_n)$ of the order of $n^{1/2}$, and preferably for the entire range values of T_n. Furthermore, we seek results that are applicable even for small samples or, more generally, for small amounts of 'observed information', and from this viewpoint the majority of results so far derived in large deviation theory are too crude in their conclusions. In other areas of application a drive for sharper results is taking place too; see, for instance, Freidlin and Wentzell (1984, section 9.1).

Formulae (6.49) and (6.51) and the supporting Theorems 6.8, 6.9, 6.11 have the degree of accuracy we are seeking.

To indicate a typical large deviation result and why it is of insufficient accuracy for our purposes consider the asymptotic behaviour as $n\to\infty$ of a mean value of the form

$$E\{e^{-nf(T_n/n)}\}$$

for some function f. In particular, if $f(x)=0$ or $+\infty$ according as $x\in B$ or $x\notin B$ this mean value reduces to $P(n^{-1}T_n\in B)$. Suppose that the normalized tilted approximation (6.51) applies to T_n so that

$$E\{e^{-nf(T_n/n)}\} \doteq nc_n \int [\det\{j_n(\hat{\lambda})\}]^{-1/2} \times \exp[-n\{f(\bar{t}) + \hat{\lambda}\cdot\bar{t} - n^{-1}k_n(\hat{\lambda})\}]d\bar{t}$$
$$\doteq nc_n \int [\det\{j_n(\hat{\lambda})\}]^{-1/2} \exp[-n\{f(\bar{t}) + n^{-1}k_n^*(\bar{t})\}]d\bar{t}, \tag{6.55}$$

where k_n^* is the Legendre transform of k_n. We now assume that $n^{-1}k_n^*$ converges to a function k^* as $n \to \infty$. This is the case in particular when the T_n are consecutive sums $S_n = X_1 + \cdots + X_n$, for then $n^{-1}k_n^* = k^*$ for every n, where k^* is the Legendre transform of the cumulant generating function for X_1. We may now, at least heuristically, apply Laplace's method to (6.55) and, taking account only of the dominating term, we are finally led to

$$-n^{-1}\log E\{e^{-nf(T_n/n)}\} \to \inf_{x\in R^m} \{f(x) + k^*(x)\}. \tag{6.56}$$

In the literature there are many variations on and generalizations of (6.56), in particular the result known as Varadhan's theorem, cf., for instance, Ellis (1985, section II.7); see also Bolthausen (1986, 1987); and formula (6.61) below.

The reason why (6.56) is not sufficiently accurate for many statistical purposes is primarily that the effect of the factors n, c_n and $[\det\{j_n(\hat{\lambda})\}]^{-1/2}$ in (6.55) has been ignored. However, this effect is, in fact, accounted for by Laplace's method which when applied more fully to (6.55) yields the approximation

$$E\{e^{-nf(T_n/n)}\} \doteq n^{1-(1/2)m}c_n[\det\{j_n(\hat{\lambda})J(\tilde{x})\}]^{-1/2}e^{-n\{f(\tilde{x})+k^*(\tilde{x})\}}, \tag{6.57}$$

where $\tilde{x}$ is the solution of

$$f(\tilde{x}) + k^*(\tilde{x}) = \inf\{f(x) + k^*(x)\}, \tag{6.58}$$

and $\tilde{\lambda}$ is the maximum likelihood estimate of λ corresponding to $\tilde{x}$, while

$$J = \nabla\nabla^{\mathrm{T}}(f + k^*). \tag{6.59}$$

Note that for the particular choice of f mentioned above, formula (6.56) specializes to

$$-n^{-1}\log P(n^{-1}T_n \in B) = \inf_{x\in B} k^*(x). \tag{6.60}$$

Some recent accounts of large deviation theory take as their starting point a generalized version of (6.60):

$$-a_n^{-1}Q_n(B) = \inf_{x \in B} k^*(x), \tag{6.61}$$

where $a_n \to \infty$ as $n \to \infty$, $\{Q_n\}$ is a sequence of probability measures, k^* is some function with values in $[0, \infty]$, and where B ranges over some class of sets $\mathscr{B}_0$. The sequence $\{Q_n\}$ is then said to satisfy a large deviation property. Under regularity conditions this allows generalization of (6.56) to

$$-a_n^{-1}\log \int e^{-a_n f(x)} Q_n(dx) \to \inf_{x \in \mathscr{X}} \{f(x) + k^*(x)\} \tag{6.62}$$

(Varadhan's theorem).

In large deviation theory and statistical mechanics the function k^* of (6.61) is termed the rate function, the entropy function or the free energy function.

6.7 Mixed tilted–direct Edgeworth expansions

It is sometimes useful to combine the techniques of sections 6.3 and 6.5 by partitioning the random vector of interest and applying an operation of partial exponential tilting, as follows.

Partition the random vector X_n as $X_n = (X_{n1}, X_{n2})$. We can write the full exponential tilt of the density $f_n(x)$, namely (6.44), as

$$f_n(X_1, X_2; \lambda_1, \lambda_2) = \exp\{\lambda_1 \cdot x_1 + \lambda_2 \cdot x_2 - k_n(\lambda_1, \lambda_2)\} f_n(x_1, x_2),$$

the partial exponential tilt then being defined as

$$f_n(x_1, x_2; \lambda_1, 0) = \exp\{\lambda_1 \cdot x_1 - k_n(\lambda_1, 0)\} f_n(x_1, x_2). \tag{6.63}$$

Note that the distribution of X_{n1} is the ordinary exponential tilt of the marginal distribution of X_{n1}.

The key equation (6.45) applies for all $\tilde{\lambda} = (\tilde{\lambda}_1, \tilde{\lambda}_2)$ and in particular to $(\tilde{\lambda}_1, 0)$, giving

$$f_n(x) = f_n(x; \tilde{\lambda}_1, 0) \exp\{k_n(\tilde{\lambda}_1, 0) - \tilde{\lambda}_1 \cdot x_1\} \tag{6.64}$$

and we proceed as before by applying an Edgeworth expansion to the first factor of (6.64). The value of $\tilde{\lambda}_1$ is to be chosen in a convenient way that will make the Edgeworth expansion as accurate as possible. A reasonable choice in general, possibly capable of improvement in particular cases, is to take $\tilde{\lambda}_1$ to be such that $E(X_{n1};\tilde{\lambda}_1,0) = x_1$, i.e. to be $\tilde{\lambda}_{10}(x_1)$, the formal maximum likelihood estimate of λ_1 under the statistical model (6.63), i.e. to satisfy

$$x_1 = [\partial k_n(\lambda_1, 0)/\partial \lambda_1]_{\lambda_1 = \hat{\lambda}_{10}}. \tag{6.65}$$

Note that $\hat{\lambda}_{10}$ is only exceptionally the same as $\hat{\lambda}_1$, the maximum likelihood estimate under the full exponential tilt.

To obtain the leading term of the expansion for $f_n(x)$, we replace the first factor of (6.64) by the leading term of the Edgeworth expansion, i.e. by the corresponding multivariate normal density. This has mean and covariance matrix respectively $(x_1, \mu_2(\hat{\lambda}_{10}, 0))$ and $\kappa(\hat{\lambda}_{10}, 0)$, where

$$\begin{aligned}\mu_2(\hat{\lambda}_{10}, 0) &= [\partial k_n(\lambda_1, \lambda_2)/\partial \lambda_2]_{\lambda_1 = \hat{\lambda}_{10}, \lambda_2 = 0},\\ \kappa(\hat{\lambda}_{10}, 0) &= [(\nabla\nabla^{\mathrm{T}} k_n)(\lambda_1, \lambda_2)]_{\lambda_1 = \hat{\lambda}_{10}, \lambda_2 = 0}.\end{aligned} \tag{6.66}$$

Thus the leading term is

$$\exp\{k_n(\hat{\lambda}_1, 0) - \hat{\lambda}_1 \cdot x_1\}\phi_m\{0, x_n - \mu_2(\hat{\lambda}_{10}, 0); \kappa(\hat{\lambda}_{10}, 0)\}, \tag{6.67}$$

where $\phi_m(x_1, x_2; \kappa)$ is the density of $\mathrm{MN}_m(0, \kappa)$.

Further terms are obtained by including additional terms of the Edgeworth series.

6.8 The delta method

In many applications exact expressions for the cumulants $\kappa^{i_1,\ldots,i_v}$ of $X = (X^1,\ldots,X^m)$, which enter the direct Edgeworth and tilted approximations discussed in sections 6.3 and 6.5, are not available in explicit form. However, in such cases it is usually possible to derive asymptotic expansions of the cumulants by a procedure known as the delta method. These expansions, developed to suitable order, may then be substituted into the Edgeworth or tilted expansion at hand, a valid and explicit asymptotic expansion for the distribution of X being thereby obtained.

To describe the method we shall assume that we are in the repeated sampling situation indicated by Theorem 6.5, and that it is desired to determine the first terms of an asymptotic expansion for the distribution of $n^{1/2}g(n^{-1/2}S_n^*)$, where g is a fixed, i.e. independent of n, and smooth function from R^m to R^d, where $d \leqslant m$, taking value zero at the origin. However, as will be obvious, the method is more widely applicable.

Thus an expansion of the Edgeworth type is sought for the probability density function $p(u)$ of the d-dimensional random variable

$$U = n^{1/2} g(n^{-1/2} S_n^*).$$

Let $g(0) = 0$ and assume that the rank of $\nabla g(0)$ is d, where ∇g is the $m \times d$ matrix of first-order partial derivatives of g. Expanding g in a Taylor series around 0 in R^m, we find

$$u^r = \sum_{\pi=1}^{\infty} \frac{1}{\pi!} g^r_{/i_1 \ldots i_\pi} x^{i_1} \cdots x^{i_\pi}. \tag{6.68}$$

This series may not be convergent but, in general, all we need is that, by virtue of $X = S_n^*$, it may be interpreted as a valid stochastic asymptotic expansion in powers of $n^{-1/2}$.

We now apply the technique for finding asymptotic expansions for cumulants of power series, set out in section 5.6, to obtain the required expressions for the cumulants.

Suppose that

$$p(s_n^*) = p^{[v]}(s_n^*) + O(n^{-(1/2)v+1})$$

as in the situation of Theorem 6.5. Then, if using the delta method we calculate the asymptotic expansions, with error terms of (absolute) order $O(n^{-(1/2)v+1})$, for the cumulants

$$\kappa^{r_1,\ldots,r_q} = K\{U^{r_1}, \ldots, U^{r_q}\}$$

of rank $q \leqslant v$, and insert these instead of the true cumulants of U in the Edgeworth approximation $p^{[v]}(u)$ for $p(u)$ we have, in great generality, that

$$p(u) = \tilde{p}^{[v]}(u) + O(n^{-(1/2)v+1}),$$

where we have used $\tilde{}$ to indicate that the exact expressions for the cumulants of U have been replaced by the asymptotic expressions discussed. This result is true, in particular, if the function g possesses continuous partial derivatives of order up to and including $v + 1$.

It is important that the method does not even require that the relevant cumulants or moments of U exist.

The delta method extends straightforwardly to more general cases where the transforming function depends on n and/or some of the components of X are discrete, provided g genuinely depends on the continuous components. We shall not discuss this in any generality but will illustrate the technique by a few examples in the Further results and exercises. For mathematically rigorous and comprehensive accounts, see Skovgaard (1981), which generalizes Bhattacharya and Ghosh (1978), and Jensen (1986, 1987).

We assumed above that the rank of $\nabla g(0)$ is d. To indicate what may happen when this is not the case, suppose $d = 1$ and $\nabla g(0) = 0$, i.e. g is a real function with value 0 and gradient 0 at 0. Then the distribution of

$$W = 2ng(n^{-(1/2)}S_n^*) \tag{6.69}$$

will, in wide generality, possess an asymptotic expansion in powers of n^{-1}, rather than $n^{-1/2}$, with leading term a χ^2-distribution on m degrees of freedom provided

$$HVH^{\mathrm{T}} = H^{\mathrm{T}} \tag{6.70}$$

where $V = \mathrm{var}(S_n^*)$ and $H = \nabla\nabla^{\mathrm{T}} g(0)$. See Chandra and Ghosh (1979), Chandra (1985) and Barndorff-Nielsen and Hall (1988). For the very simplest case, see Example 2.9.

We have seen how to generalize the delta method of local linearization from functions of one variable to functions of several or indeed many variables. Especially in semiparametric and nonparametric statistics, it is helpful to have a further generalization in which, in particular, the quantity under study is a functional, i.e. a real-valued function whose argument is itself a function. Thus we may have a real number associated with each member of some space of distribution functions, i.e. a mapping from some space of distribution functions to the real line.

The most satisfactory way of dealing with such problems is to extend the notion of differentiation; there are several ways of doing this which differ essentially only in the regularity conditions imposed.

The most familiar elementary example of the argument involved is probably the derivation of Euler's equation in the calculus of variations and we therefore motivate the general definition and notation via that; some readers may prefer to go straight to the general definition.

We consider a family $\mathscr{G}$ of differentiable functions g of a real variable t representing curves joining two points (a_0, b_0) and (a_1, b_1), i.e. $g(a_i) = b_i$ $(i = 0, 1)$. For each such function let there be defined a real number $\phi(g)$ given by

$$\phi(g) = \int_{a_0}^{a_1} q\{t, g(t), \dot{g}(t)\}\, dt,$$

where q is a given function and $\dot{g} = dg/dt$. This is a mapping from g to the real line.

Suppose now that g is perturbed slightly to $g + \varepsilon h$, where h vanishes at a_0 and a_1. Then q changes to

$$q\{t, g(t), \dot{g}(t)\} + \varepsilon\{(\partial q/\partial g)h + (\partial q/\partial \dot{g})\dot{h}\} + o(\varepsilon).$$

On integrating the term involving $\dot{h}$ by parts, we have that

$$\phi(g + \varepsilon h) = \phi(g) + \int \{\partial q/\partial g - (d/dt)\partial q/\partial \dot{g}\}\varepsilon h\, dt + o(\varepsilon). \quad (6.71)$$

It is now natural to regard the correction term as the result of a generalized notion of differentiation applied to ϕ, rewriting the equation in the form

$$\phi(g + \varepsilon h) = \phi(g) + \dot{\phi}(g)(\varepsilon h) + o(\varepsilon) \quad (6.72)$$

where $\dot{\phi}(g)$ applied to εh represents the operation of forming the integral in (6.71). The way that this is used in the calculus of variations is to take the vanishing of $\dot{\phi}$ as the condition for g to give a stationary value to ϕ; note that the value of the expression in parentheses in (6.71) at a particular t gives the influence of a perturbation localized near t.

The distinction between a number of ideas of generalized differentiation lies in the precise meaning of the term $o(\varepsilon)$ in (6.72). In general we have a space of elements g for each of which is defined a real number $\phi(g)$. If the space is equipped with a norm we take the existence of a continuous linear functional $\dot{\phi}$ such that (6.72) holds as defining the Fréchet differentiability of ϕ, the term $o(\varepsilon)$ meaning a quantity that tends to zero more rapidly than the norm of εh. A weaker and hence more general notion is that of Hadamard differentiability. Here only topological requirements are placed on $\mathscr{G}$, that is the existence of open, closed and compact sets, and (6.72) is required to hold uniformly for h in any compact set in $\mathscr{G}$ as $\varepsilon \to 0$.

Extensions are possible to repeated differentiation and to vector-valued ϕ and also some weakening of the key regularity assumptions is allowable.

In the simplest statistical application of these ideas the space, in general now denoted by $\mathscr{F}$, is the space of distribution functions, F, on the real line and the functional ϕ, often called a *statistical* or *von Mises functional*, has the form, for some function r,

$$\phi(F) = \int r(t, F)dF(t). \tag{6.73}$$

It will often be the case that the Hadamard derivative of ϕ has the form

$$\dot{\phi}(F)(\varepsilon h) = \varepsilon \int \psi(t; F; \phi)dh(t).$$

The curve of ψ against t is called the *influence curve*; it indicates the sensitivity of ϕ to perturbations of F near t.

Example 6.4 Mean and variance as functionals

If

$$\phi(F) = \int t\, dF(t),$$

the mean value of F, then

$$\phi(F + \varepsilon h) = \phi(F) + \varepsilon \int t\, dh(t)$$

and hence $\psi(t; F; \phi)$ is simply t.

Next, suppose

$$\phi(F) = \int \{t - \mu(F)\}^2 dF(t)$$

where $\mu(F) = \int t\, dF(t)$. Then

$$\phi(F + \varepsilon h) = \phi(F) + \varepsilon\dot{\phi}(F)(h) + \tfrac{1}{2}\varepsilon^2 \ddot{\phi}(F)(h),$$

where

$$\dot{\phi}(F)(h) = \int \{t - \mu(F)\}^2 dh(t),$$

so that

$$\psi(t; F; \phi) = \{t - \mu(F)\}^2.$$

Furthermore,

$$\ddot{\phi}(F)(h) = -2\left\{\int t\, dh(t)\right\}^2.$$

Let $Y_1, \ldots, Y_n$ be independent and identically distributed random variables, copies of a random variable Y with distribution function F. Let F_n be the sample distribution function. Then approximately

$$\phi(F_n) - \phi(F) = \int \psi(t; F; \phi) d(F_n - F).$$

It can be shown that in wide generality $\phi(F_n)$ is asymptotically normal with mean $\phi(F)$ and asymptotic variance

$$n^{-1} \int \psi^2(x; F; \phi) dF(x). \tag{6.74}$$

Suppose in particular that $\phi(F)$ satisfies the equation

$$\int g\{\phi(F); t\} dF(t) = 0$$

for some function g of two real variables. Then under suitable regularity conditions

$$\psi(t; F; \phi) = -g\{\phi(F), t\} \Big/ \int \dot{g}\{\phi(F); t\} dF(t),$$

where $\dot{g}(\theta; t) = \partial g(\theta; t)/\partial\theta$.

Here, with $Y_1, \ldots, Y_n$ as above, g determines an unbiased estimating equation for θ in the form

$$\sum g(\theta; Y_i) = 0.$$

Such equations can, however, be studied by more elementary methods; cf. Exercise 4.16.

For details and further examples, see Reeds (1976), Serfling (1980), Reid (1983), Withers (1983), Fernholz (1988), Gill (1987) and the references provided there.

Extension of the delta method leads to expansions for random variables and stochastic processes of considerable generality. The next example is an illustration.

Example 6.5 Expansions of nonlinear stochastic processes

In discrete time many interesting models, especially for stationary processes, can be formulated in terms of a sequence $\{Z_t\}$ of independent and identically distributed random variables of zero mean. In particular we have autoregressive representations in which the process $\{Y_t\}$ is constructed with $\{Z_t\}$ as an innovation process

$$Y_t = h(Y_{t-1}, Y_{t-2}, \ldots; Z_t) \tag{6.75}$$

and moving average representations in which Y_t is defined directly in terms of the 'pure noise' sequence $\{Z_t\}$

$$Y_t = g(Z_t, Z_{t-1}, \ldots). \tag{6.76}$$

The special cases in which the functions h and g are linear, and $\{Z_t\}$ normal, are the most familiar and widely studied, the relation between (6.75) and (6.76) then being well understood.

A wide class of nonlinear processes can be constructed formally by taking the functions h and g to be nonlinear; a natural possibility is to write

$$Y_t = \sum \psi_i Z_{t-i} + \sum_{i \neq j} \psi_{ij} Z_{t-i} Z_{t-j} + \cdots. \tag{6.77}$$

It follows formally that if, without loss of generality, we take $E(Z_t^2) = 1$, then

$$\psi_i = E(Y_t Z_{t-i}), \qquad \psi_{ij} = E(Y_t Z_{t-i} Z_{t-j}) \qquad (i \neq j), \tag{6.78}$$

etc. One possibility is to take the higher-degree terms as a small perturbation of the linear terms and to use this as a basis for an asymptotic expansion.

Note that if we are interested in some function of the process of interest, e.g. $l(Y_t, Y_{t-1}, \ldots)$, formally the same kind of expansion will apply.

Further, if Z_t is normally distributed we may be able to simplify the problem by an orthonormal transformation to

$$\tilde{Z}_1 = \sum \phi_{1j} Z_{t-j}, \qquad \tilde{Z}_2 = \sum \phi_{2j} Z_{t-j}, \ldots$$

and by introducing the Hermite polynomials $U_{rs} = H_r(\tilde{Z}_s)$.

These are orthogonal random variables and a general polynomial expansion (6.77) can be re-expressed as a weighted sum of products of the U_{rs}. Suitable choice of the ϕ_{ij} can simplify particular problems.

When we pass to continuous time, serious technical issues arise.

The sequence $\{Z_t\}$ is replaced by a unit Brownian motion $\{B(t)\}$ and (6.75) corresponds in simple cases to a stochastic differential equation. It can be shown (Itô, 1951) that any (square integrable) functional $\Psi[Y(\cdot)]$ of a Gaussian process $\{Y(t):0 \leqslant t \leqslant T\}$ has an expansion analogous to (6.77),

$$\Psi[Y(\cdot)] = \int_0^T \psi_1(u)dB(u) + \int_0^T \int_{-\infty}^{u_1} \psi_2(u_1, u_2)dB(u_1)dB(u_2) + \cdots, \tag{6.79}$$

where the stochastic integrals are to be interpreted in the sense of Itô. In analogy with (6.78), we have that

$$E\{\Psi[Y(\cdot)]dB(u_1)dB(u_2)\} = \psi_2(u_1, u_2)du_1 du_2,$$

etc. Further it can be shown, by introducing a complete orthonormal system ϕ_r of square integrable functions on the real interval over which the process is defined and then writing

$$U_{rs} = H_r\left(\int \phi_s(u)dB(u)\right), \tag{6.80}$$

that the expansion (6.79) can be expressed in terms of a weighted series of products of the $\{U_{rs}\}$.

Two very special cases have stochastic differential equations corresponding to (6.75) of the form

$$dY(t) = -\alpha Y(t)dt + \gamma\, dB(t), \tag{6.81}$$

the Ornstein–Uhlenbeck process in which we suppose $Y(0) = \xi_0$, $\alpha > 0$, $\gamma > 0$, and the form

$$dY(t) = \gamma f(t) Y(t) dB(t), \tag{6.82}$$

where $Y(0) > 0$, $\gamma > 0$ and $f(t)$ is a known deterministic function of t. Suppose further that $\Psi[Y(\cdot)]$ is simply $Y(t)$, the value of $Y(\cdot)$ at a particular fixed time point t.

For (6.81) it can be shown that the expansion (6.79) takes the anticipated form

$$Y(t) = \xi_0 e^{-\alpha t} + \gamma \int_0^t e^{-\alpha(t-s)} dB(s), \tag{6.83}$$

whereas for (6.82)

$$Y(t) = \xi_0 \left\{ 1 + \sum_{n=1}^{\infty} \gamma^n \int_0^t \int_0^{t_{n-1}} \cdots \int_0^{t_1} f(u_1) \cdots f(u_n) dB(u_1) \cdots dB(u_n) \right\}$$
$$= \xi_0 \exp\left\{ \gamma \int_0^t f(s) dB(s) - \tfrac{1}{2}\gamma^2 \int_0^t \{f(s)\}^2 dB(s) \right\}, \tag{6.84}$$

so that in particular $Y(t)$ has a log normal distribution.

For details, see Hida (1980) and Isobe and Sato (1983).

6.9 Generalized formal Edgeworth expansions

Suppose we wish to derive an expansion of a probability function $f(x)$, as a series with leading term $f^\dagger(x)$, where $f^\dagger(x)$ is some other probability function. Let M, $M^\dagger$, K and $K^\dagger$ denote the corresponding moment generating and cumulant generating functions, the associated moments and cumulants being denoted by κ_1^I, $\kappa_1^{\dagger I}$, κ_0^I and $\kappa_0^{\dagger I}$.

Writing

$$\lambda_0^I = \kappa_0^I - \kappa_0^{\dagger I},$$

we then have

$$K(t) = K^\dagger(t) + \sum_{|I|=1}^{\infty} \frac{1}{|I|!} \lambda_0^I t_I \tag{6.85}$$

and

$$M(t) = M^\dagger(t) \left\{ 1 + \sum_{|I|=1}^{\infty} \frac{1}{|I|!} \lambda_1^I t_I \right\}, \tag{6.86}$$

where

$$\lambda_1^I = \sum_{\alpha=1}^{|I|} \sum_{I/\alpha} \lambda_0^{I_1} \cdots \lambda_0^{I_\alpha},$$

cf. (5.27).

Inversion of (6.86) gives, on account of section 6.11,

$$f(x) = f^\dagger(x) \left\{ 1 + \sum_{|I|=1}^{\infty} \frac{1}{|I|!} \lambda_1^I h_I^\dagger(x) \right\}, \tag{6.87}$$

where

$$h_I^\dagger(x) = (-1)^{|I|} \{ f^\dagger(x) \}^{-1} f^{\dagger(I)}(x). \tag{6.88}$$

Note that in general it will be necessary to rearrange the terms in (6.87), i.e. to collect terms of the same asymptotic order, in order to obtain a valid asymptotic expansion for $f(x)$.

To avoid the problem that the right-hand side of (6.87) may be negative for certain values of x we may, again formally, exponentiate the quantity in curly brackets to obtain

$$f(x) = f^{\dagger}(x) \exp\left\{ \sum_{|I|=1}^{\infty} \frac{1}{|I|!} \sum_{\alpha=1}^{|I|} \sum_{I/\alpha} \lambda_0^{I_1} \cdots \lambda_0^{I_\alpha} h_I^{\dagger}(x) \right\}, \tag{6.89}$$

where $h_{I_*}^{\dagger}$ is defined, by analogy with the generalized c-coefficients (5.32), by

$$h_{I_*}^{\dagger}(x) = \sum_{I_* \vee I'_* = I} h_{0I'_1}^{\dagger}(x) \cdots h_{0I'_\beta}^{\dagger}(x), \tag{6.90}$$

where, in turn, $h_{0I}^{\dagger}(x)$ is defined as

$$h_{0I}^{\dagger} = \sum_{\alpha=1}^{|I|} (-1)^{\alpha-1}(\alpha-1)! \sum_{I/\alpha} h_{I_1}^{\dagger}(x) \cdots h_{I_\alpha}^{\dagger}(x). \tag{6.91}$$

Note, however, the problems about convergence discussed in the univariate case in section 4.2, point 4.

6.10 Inversion

We briefly discuss the multivariate versions of two types of inversion problem that occur rather frequently in asymptotic calculations. One is that of expressing the higher-order derivatives of the inverse of an m-dimensional one-to-one mapping in terms of the higher-order derivatives of the mapping itself; the solution is a basis for a generalization of Lagrange's formula. For a statement and proof of the latter, see, for instance, Dieudonné (1968). The other problem is that of inverting a commonly occurring type of multivariate asymptotic expansion. The one-dimensional case of the latter problem was treated in section 3.5. For proofs of the statements below, see Barndorff-Nielsen and Blæsild (1988b).

We shall repeatedly refer to the following formula for the inverse of a lower triangular $m \times m$ matrix with unit diagonal. Let a_j^i constitute such a matrix, i.e.

$$a_j^i = \begin{cases} 0 & \text{if } i > j, \\ 1 & \text{if } i = j, \\ \text{arbitrary real} & \text{if } i < j, \end{cases}$$

and let b^i_j be the inverse matrix, which is again a lower triangular matrix with unit diagonal. Then for $i<j$, b^i_j may be expressed as

$$b^i_j = -a^i_j + \sum_{\pi=2}^{j-i} (-1)^\pi \sum{}^* a^i_{k_1} a^{k_1}_{k_2} \cdots a^{k_{\pi-2}}_{k_{\pi-1}} a^{k_{\pi-1}}_j, \tag{6.92}$$

where the second term on the right-hand side is interpreted as 0 if $i=j+1$ and where otherwise the inner sum is over all $k_1,\ldots,k_{\pi-1}$ such that

$$i<k_1<\cdots<k_{\pi-1}<j.$$

Let $u=(u^1,\ldots,u^m)$ be a one-to-one and smooth function of $x=(x^1,\ldots,x^m)$, where x varies over some domain $D\subset R^m$, and denote generic coordinates of u and x by $u^r, u^s, u^t,\ldots$ and $x^i, x^j, x^k,\ldots$, respectively. Let $u^r_{/i_1\cdots i_\nu}$ denote the νth order derivative of u^r with respect to $x^{i_1},\ldots,x^{i_\nu}$, etc., and, more generally, with $I_\mu=i_1\cdots i_\mu$, $J_\nu=j_1\cdots j_\nu$, $R_\mu=r_1\cdots r_\mu$, $S_\nu=s_1\cdots s_\nu$ and, for $\mu\leqslant\nu$, write

$$u^{R_\mu}_{/J_\nu} = \sum_{J_\nu/\mu}^{\rightarrow} u^{r_1}_{/J_{\nu 1}} \cdots u^{r_\mu}_{/J_{\nu\mu}}, \qquad x^{I_\mu}_{/S_\nu} = \sum_{S_\nu/\mu}^{\rightarrow} x^{i_1}_{/S_{\nu 1}} \cdots x^{i_\mu}_{/S_{\nu\mu}}. \tag{6.93}$$

The arrows indicate that the summation is over *ordered partitions*, where a partition $J_{\nu 1},\ldots,J_{\nu\mu}$ of J_ν into μ blocks is *ordered* if (i) the order of the indices within each block is the same as their order within J_ν, and (ii) the first index in $J_{\nu\lambda}$ comes before the first index in $J_{\nu,\lambda+1}$, $\lambda=1,\ldots,\mu-1$ as compared to the ordering within J_ν. Finally, for $\mu>\nu$ we interpret the quantities in (6.93) as 0.

The notation (6.93) allows us, in particular, to express compactly the derivatives of a real function f of m variables $x^1,\ldots,x^m$ in terms of the derivatives of f when f is considered as a function of an alternative set of coordinates $u^1,\ldots,u^m$. With our notational conventions the derivatives in question are $f_{/I_\nu}$ and $f_{/R_\nu}$, and we have

$$f_{/I_\nu} = \sum_{\lambda=1}^{\nu} f_{/R_\lambda} u^{R_\lambda}_{/I_\nu}, \tag{6.94}$$

as may be shown by induction.

Since the identity mapping may be expressed as the composed mapping $x\to u\to x$ we have

$$x^i_{/r} u^r_{/j} = \delta^i_j, \tag{6.95}$$

where the right-hand side is the Kronecker delta. Repeated

differentiation of (6.95) yields, for $1 \leqslant \lambda \leqslant \mu$,

$$\sum_{\nu=\lambda}^{\mu} x^{I_\lambda}_{/S_\nu} u^{S_\nu}_{/J_\mu} = \delta^{I_\lambda}_{J_\mu}, \tag{6.96}$$

where the right-hand side is again a Kronecker delta:

$$\delta^{I_\lambda}_{J_\mu} = \begin{cases} 1 & \text{if } \lambda = \mu \text{ and } I_\lambda = J_\mu, \\ 0 & \text{otherwise.} \end{cases}$$

Note that, in particular, for $\mu = \nu$,

$$u^{R_\mu}_{/J_\mu} = u^{r_1}_{/j_1} \cdots u^{r_\mu}_{/j_\mu}, \qquad x^{I_\mu}_{/S_\mu} = x^{i_1}_{/s_1} \cdots x^{i_\mu}_{/s_\mu}.$$

Let

$$u^{I_\mu}_{/J_\nu} = x^{I_\mu}_{/R_\mu} u^{R_\mu}_{/J_\nu}, \qquad x^{R_\mu}_{/S_\nu} = u^{R_\mu}_{/I_\mu} x^{I_\mu}_{/S_\nu}, \tag{6.97}$$

$$u^{R_\mu}_{/S_\nu} = x^{I_\nu}_{/S_\nu} u^{R_\mu}_{/I_\nu}, \qquad x^{I_\mu}_{/J_\nu} = u^{R_\nu}_{/J_\nu} x^{I_\mu}_{/R_\nu}. \tag{6.98}$$

Using (6.97)–(6.98) we may rewrite (6.96) as

$$\sum_{\nu=\lambda}^{\mu} x^{R_\lambda}_{/T_\nu} u^{T_\nu}_{/S_\mu} = \delta^{R_\lambda}_{S_\mu} \tag{6.99}$$

or, dually, as

$$\sum_{\nu=\lambda}^{\mu} x^{I_\lambda}_{/K_\nu} u^{K_\nu}_{/J_\mu} = \delta^{I_\lambda}_{J_\mu}. \tag{6.100}$$

The point of introducing (6.97)–(6.98) is that for any given $n = 1, 2, \ldots$, the multiarrays $u^{I_\mu}_{/J_\nu}$, $1 \leqslant \mu$, $\nu \leqslant n$, on the one hand, and the multiarrays $x^{I_\mu}_{/J_\nu}$, $1 \leqslant \mu$, $\nu \leqslant n$, on the other, each constitute a lower triangular matrix with unit diagonal, and similarly for $x^{R_\mu}_{/S_\nu}$ and $u^{R_\mu}_{/S_\nu}$. On account of (6.100) the first two matrices are mutually inverse, and similarly for the two latter matrices. We may therefore apply formula (6.92) to express $u^{I_\mu}_{/J_\nu}$ in terms of $x^{I_\mu}_{/J_\nu}$. Specifically we have for $1 \leqslant \mu < \nu$,

$$u^{I_\mu}_{/J_\nu} = -x^{I_\mu}_{/J_\nu} + \sum_{\pi=2}^{\nu-\mu} (-1)^\pi \sum{}^{*}\, x^{I_\mu}_{/K_{(1)}} x^{K_{(1)}}_{/K_{(2)}} \cdots x^{K_{(\pi-2)}}_{/K_{(\pi-1)}} x^{K_{(\pi-1)}}_{/J_\nu}, \tag{6.101}$$

the sum being over all $K_{(1)}, \ldots, K_{(\pi-1)}$ such that

$$\mu < |K_{(1)}| < \cdots < |K_{(\pi-1)}| < \nu;$$

for $\nu - \mu = 1$ the sum is interpreted as 0.

Formula (6.101) is the desired formula for expressing not only the derivatives of u with respect to the coordinates of x but also the associated quantities $u^{R_\mu}_{/J_\nu}$ in terms of the analogous quantities for the mapping from u to x. We emphasize again the step from $u^{R_\mu}_{/J_\nu}$ to $u^{I_\mu}_{/J_\nu}$ by multiplication of the former with $x^{I_\lambda}_{/R_\lambda}$, etc. (definitions (6.97)–(6.98)). Thus there is the implicit assumption in the above that $x^i_{/r}$ is tractable, though $x^i_{/R_\nu}$ is not directly so for $\nu > 1$. However, $x^i_{/r}$ is of course just the inverse of $u^r_{/i}$.

The expression for $u^r_{/I_\nu}$ determined by (6.101) is, for $\nu = 2, 3$ and 4,

$$
\begin{aligned}
u^r_{/j_1j_2} &= -u^{rs_1s_2}_{/ij_1j_2} x^i_{/s_1s_2}, \\
u^r_{/j_1j_2j_3} &= -u^{rs_1s_2s_3}_{/ij_1j_2j_3}\{x^i_{/s_1s_2s_3} - u^t_{/k} x^i_{/s_1t} x^k_{/s_2s_3}[3]\}, \\
u^r_{/j_1j_2j_3j_4} &= -u^{rs_1s_2s_3s_4}_{/ij_1j_2j_3j_4}\{x^i_{/s_1s_2s_3s_4} - u^t_{/k}(x^i_{/s_1t} x^k_{/s_2s_3s_4}[4] \\
&\quad + x^i_{/s_1s_2t} x^k_{/s_3s_4}[6]) + u^{tu}_{/kl}(x^i_{/tu} x^k_{/s_1s_2} x^l_{/s_3s_4}[3] + x^i_{/s_1t} x^k_{/s_2u} x^l_{/s_3s_4}[12]\}.
\end{aligned}
\tag{6.102}
$$

As an important type of application, suppose f is a real function defined on a neighbourhood of 0 in R^m and that we wish to determine the Taylor expansion around 0 of f in powers of x. Suppose further that both f and x are analytically tractable as functions of u, whereas as functions of x both f and u are more or less unmanageable by direct calculation. The problem is then to find $f_{/I_\nu}$ in terms of the quantities $f_{/R_\lambda}$ and $x^i_{/R_\lambda}$, and this is achieved by the two formulae (6.94) and (6.101).

Example 6.6 Exponential family

Let $k = k(\theta)$ be the cumulant transform of a core exponential family and let $\tau = \tau(\theta)$ be the mean value parameter (cf. Theorem 6.1). The derivatives of k with respect to θ, i.e.

$$k_{/I_\nu}(\theta) = K\{S_{i_1}, \ldots, S_{i_\nu}\} = \kappa_{i_1,\ldots,i_\nu}$$

are the cumulants of the canonical statistic $S = (S_1, \ldots, S_m)$. Here we seek the derivatives of k with respect to the coordinates of τ, when k is considered as a function of τ.

Let $\kappa^{i,j}$ denote the inverse matrix of $\kappa_{i,j} = \mathrm{cov}(S_i, S_j)$, and let $S^i = \kappa^{i,i'} S_{i'}$ and

$$\kappa^{i_1,\ldots,i_\nu} = K\{S^{i_1}, \ldots, S^{i_\nu}\},$$

the elemental cumulant of $S^{i_1}, \ldots, S^{i_\nu}$. By (6.102) the first derivatives of k with respect to τ may then be written as

$$\begin{aligned} \partial k/\partial\tau_i &= -\kappa^{i,j}\tau_j, \qquad \partial^2 k/\partial\tau_i\partial\tau_j = -\kappa^{i,j,k}\tau_k + \kappa^{i,j}, \\ \partial^3 k/\partial\tau_i\partial\tau_j\partial\tau_k &= -2\kappa^{i,j,k} - (\kappa^{i,j,k,l} - \kappa^{i,j,m}\kappa^{k,l,n}\kappa_{m,n}[3])\tau_l. \end{aligned} \tag{6.103}$$

We now turn to the second of the two inversion problems indicated at the outset. Let $z = (z^1,\ldots,z^m)$ and $w = (w^1,\ldots,w^m)$ be connected by m relations

$$w^i = \sum_{\nu=1}^{\infty} a^i_{J_\nu} z^{J_\nu} \tag{6.104}$$

($i,j = 1,\ldots,m$ and $J_\nu = j_1\cdots j_\nu$) which may be thought of as a system of m formal power series. In particular, the series could be ordinary power series or asymptotic series with the coefficients $a^i_{J_\nu}$ of order $O(1)$ and with z^i of order $O(n^{-1/2})$, and hence z^{I_ν} of order $O(n^{-(1/2)\nu})$, for some parameter $n \to \infty$. For specificity, we focus on this latter case. Further, for simplicity, we assume that the $a^i_{J_\nu}$ are all symmetric in the lower indices and that $a^i_j = \delta^i_j$, as can usually be arranged. We wish to invert (6.104), i.e. to find an asymptotic expansion of z^i ($i = 1,\ldots,m$) of the form

$$z^i = \sum_{\nu=1}^{\infty} b^i_{J_\nu} w^{J_\nu}, \tag{6.105}$$

as determined by (6.104). In principle, the technique for doing this is similar to that for the case $m = 1$ discussed in section 3.5. To handle the multivariate case we introduce the notation

$$a^{I_\mu}_{J_\nu} = \sum_{J_\nu/\mu}{}' a^{i_1}_{J_{\nu 1}}\cdots a^{i_\mu}_{J_{\nu\mu}}, \qquad b^{I_\mu}_{J_\nu} = \sum_{J_\nu/\mu}{}' b^{i_1}_{J_{\nu 1}}\cdots b^{i_\mu}_{J_{\nu\mu}}, \tag{6.106}$$

where the summations are as in (6.93) except that the partitions $I_{\nu 1},\ldots,I_{\nu\mu}$ have to be *strictly ordered*, i.e. the order in which the indices i occur in the combined sequence $I_{\nu 1}\cdots I_{\nu\mu}$ is the same as their order in I_ν. This is for $\mu \leqslant \nu$, while for $\mu > \nu$ $a^{T_\mu}_{J_\nu}$ and $b^{I_\mu}_{J_\nu}$ are taken to be 0.

For any given natural number n, the two collections of arrays $a^{I_\mu}_{J_\nu}$, $1 \leqslant \mu$, $\nu \leqslant n$, and $b^{I_\mu}_{J_\nu}$, $1 \leqslant \mu$, $\nu \leqslant n$, each constitute a lower triangular matrix with unit diagonal, and these matrices, on account of the definition of (6.105) as the inversion of (6.104), can be shown to be each other's inverse. Hence we may apply formula (6.92) to obtain

$$b_{J_\nu}^{I_\mu} = -a_{J_\nu}^{I_\mu} + \sum_{\pi=2}^{\nu-\mu} (-1)^\pi \sum{}^* a_{K_{(1)}}^{I_\mu} a_{K_{(2)}}^{K_{(1)}} \cdots a_{K_{(\pi-1)}}^{K_{(\pi-2)}} a_{J_\nu}^{K_{(\pi-1)}}, \quad (6.107)$$

the summation conventions being as in (6.101).

In particular, taking $\mu = 1$, we find that the first few terms of (6.105) are given by

$$\begin{aligned} z^i = w^i &- a^i_{j_1 j_2} w^{j_1 j_2} \\ &- (a^i_{j_1 j_2 j_3} - 2a^i_{j_1 k} a^k_{j_2 j_3}) w^{j_1 j_2 j_3} \\ &- (a^i_{j_1 j_2 j_3 j_4} - 2a^i_{j_1 k} a^k_{j_2 j_3 j_4} - 3a^i_{j_1 j_2 k} a^k_{j_3 j_4} \\ &+ a^i_{kl} a^k_{j_1 j_2} a^l_{j_3 j_4} + 4a^i_{j_1 k} a^k_{j_2 l} a^l_{j_3 j_4}) w^{j_1 j_2 j_3 j_4} + \cdots. \end{aligned} \quad (6.108)$$

6.11 Appendix: Fourier transformation

We now collate a number of standard results on Fourier transforms. Let f denote a function (possibly complex-valued) on R^m and suppose that $f \in L^p(R^m)$ for some $p \geqslant 1$, i.e.

$$\int |f(x)|^p dx < \infty .$$

The *Fourier transform* of f is the function $\vec{f}$ defined, also on R^m, by

$$\vec{f}(\zeta) = \int e^{i\zeta \cdot x} f(x) dx, \quad (6.109)$$

where i is the imaginary unit. The *inverse Fourier transform* $\overleftarrow{f}$ of f is defined by

$$\overleftarrow{f}(\zeta) = (2\pi)^{-m} \int e^{-i\zeta \cdot x} f(x) dx. \quad (6.110)$$

If $f \in L^1$ then $\vec{f}$ is bounded and uniformly continuous on R^m.

On the assumption that $\vec{f} \in L^1$ we thus have that the function $\overleftrightarrows{f}$ given by

$$\overleftrightarrows{f}(x) = (2\pi)^{-m} \int e^{-ix\cdot\zeta} \vec{f}(\zeta) d\zeta$$

is bounded and uniformly continuous and $\overleftrightarrows{f}$ is almost everywhere equal to $f(x)$.

Let x^r and ζ_r denote the generic coordinates of x and ζ, respectively ($r = 1,\ldots,m$). With $R = r_1 \ldots . r_\nu$ an arbitrary index set we write $|R| = \nu$, $x^R = x^{r_1} \cdots x^{r_\nu}$, $\zeta_R = \zeta_{r_1} \cdots \zeta_{r_\nu}$, and we let $f^{(R)}(x)$ and $g^{(R)}(\zeta)$ denote the corresponding partial derivatives of the functions $f(x)$ and $g(\zeta)$.

Formal differentiation of

$$f(x) = (2\pi)^{-m} \int e^{-ix\cdot\zeta} \vec{f}(\zeta) d\zeta \tag{6.111}$$

yields

$$f^{(R)}(x) = (2\pi)^{-m} (-i)^{|R|} \int e^{-ix\cdot\zeta} \zeta_R \vec{f}(\zeta) d\zeta, \tag{6.112}$$

showing that $f^{(R)}(x)$ is the inverse Fourier transform of

$$(-i)^{|R|} \zeta_R \vec{f}(\zeta). \tag{6.113}$$

The precise result is that if for a given positive integer ν (6.113) belongs to L^1 for every index set R with $|R| = \nu$, then a version of f is ν times differentiable with νth order derivatives expressible via (6.112).

Conversely, if the function g defined by

$$g(x) = x^R f(x) \tag{6.114}$$

belongs to L^1, then

$$\vec{g}(\zeta) = (-i)^{|R|} \vec{f}^{(R)}(\zeta). \tag{6.115}$$

On the whole, the above results extend to the case where instead of the measure $f(x)dx$ in (6.109) one substitutes an arbitrary signed finite measure P on R^m, the resulting (Fourier–Stieltjes) transform being

$$\phi(\zeta) = \int e^{i\zeta\cdot x} dP. \tag{6.116}$$

When P is a probability measure, as we shall henceforth assume, (6.116) is, of course, the characteristic function of P.

Again, if $\phi \in L^1$ then there exists a bounded and uniformly continuous f such that $dP(x) = f(x)dx$.

Further, in the present more general framework the role of the inversion formula is taken over by an integrated version of (6.110). Specifically, letting $X = (X^1, \ldots, X^m)$ denote the random variable whose distribution is given by P, we have

$$P(a < X \leqslant b) = \lim_{c\to\infty} (2\pi)^{-m} \int_{-c}^{c} \cdots \int_{-c}^{c} \prod_{k=1}^{m} \frac{e^{i\zeta_k a_k} - e^{i\zeta_k b_k}}{i\zeta_k} \phi(\zeta) d\zeta \quad (6.117)$$

for all a, $b \in R^m$ such that $a \leqslant b$ and the probability mass of the boundary of the m-dimensional interval determined by a and b is 0.

In perfect extension of (6.112)–(6.113) we have that if ϕ, given by (6.116), is such that

$$(-1)^{|R|} \zeta_R \phi(\zeta) \quad (6.118)$$

is integrable then P has a probability density function f such that (6.118) is the Fourier transform of $f^{(R)}(x)$.

Suppose $E(|X^r|^{n+\delta}) < \infty$ for $r = 1, \ldots, m$ and some $\delta > 0$. Then $\phi^{(R)}$ exists and is bounded and uniformly continuous for every $R = r_1 \ldots r_\nu$ with $\nu \leqslant n$, and

$$\phi^{(R)}(\zeta) = i^\nu \int x^R e^{i\zeta \cdot x} dP. \quad (6.119)$$

Moreover,

$$\phi(\zeta) = \sum_{\nu=0}^{n-1} \frac{i^\nu}{\nu!} \kappa^{r_1 \cdots r_\nu} \zeta_{r_1} \cdots \zeta_{r_\nu} + \rho_n(\zeta), \quad (6.120)$$

where

$$|\rho_n(\zeta)| \leqslant \frac{1}{n!} (|\zeta_1| + \cdots + |\zeta_m|)^n \max_r E(|X^r|^n). \quad (6.121)$$

This follows, by use of Hölder's inequality, from the following version of Taylor's theorem.

Let f be a complex-valued function defined on an open interval J of the real line and having continuous derivatives of orders $\nu = 1, \ldots, n$. If x, $x + h \in J$ then

$$f(x+h) = f(x) + \sum_{\nu=1}^{n-1} \frac{h^\nu}{\nu!} f^{(\nu)}(x) + \frac{h^n}{(n-1)!} \int_0^1 (1-u)^{n-1} f^{(n)}(x+uh) du. \quad (6.122)$$

Finally, we reconsider briefly (6.112) which in the one-dimensional case can be rewritten as

$$f^{(\nu)}(x) = (2\pi)^{-1} (-1)^\nu \int e^{-ix\zeta} (i\zeta)^\nu \vec{f}(\zeta) d\zeta.$$

This suggests extending the concept of derivatives of f by allowing ν to take any real value for which the integral exists. This leads to a definition of so-called *derivatives of fractional order* different from but essentially equivalent to the classical definition of Liouville. We have used the idea of such derivatives briefly in our discussion of Edgeworth-like expansions for stable laws, particularly in (4.30).

6.12 Appendix: Laplace's method: regularity conditions

Theorem 6.2

Let f and g be real-valued functions defined on a subset D of R^m such that:

(i) f has an absolute maximum value at an interior point ξ of D and $f(\xi) > 0$;
(ii) there exists a constant $s > 0$ such that gf^s is absolutely integrable on D;
(iii) all partial derivatives $\partial f/\partial x_i$ and $\partial^2 f/\partial x_i \partial x_j$ exist and are continuous in a neighbourhood N of ξ;
(iv) there exists a constant $A < 1$ such that $|f(x)/f(\xi)| < A$ for all $x \in D - N$;
(v) g is continuous in a neighbourhood of ξ and $g(\xi) \neq 0$.

Then for $n \to \infty$

$$\int_D g(x)\{f(x)\}^n dx \sim g(\xi)(2\pi/n)^{(1/2)m}\{\Delta(\xi)\}^{-1/2}\{f(\xi)\}^n, \tag{6.123}$$

where $\Delta(\xi)$ is the Hessian determinant of $-\log f(x)$ evaluated at $x = \xi$, i.e.

$$\Delta(x) = -\det\{\nabla\nabla^{\mathrm{T}} \log f(x)\}. \tag{6.124}$$

To derive this result, write the integral in (6.123) as

$$\{f(\xi)\}^n \int_D g(x) \exp[n\{\log f(x) - \log f(\xi)\}]dx$$

and choose a neighbourhood N of 0 such that $\log f(x) - \log f(\xi)$ is approximately equal to

$$-\tfrac{1}{2}(x-\xi)^{\mathrm{T}}\Delta(\xi)(x-\xi)$$

for $x - \xi \in N$. Then take n sufficiently large to make the integral over $-\xi + D - N$ negligible. See Hsu (1948) for details.

6.13 Appendix: Direct and tilted Edgeworth expansions: regularity conditions

Except at the very end of this section the random variables considered will, by the conditions to be imposed, have non-lattice distributions.

Throughout the section $Y_1, Y_2, \ldots$ denote independent random vectors of dimension m and with mean 0 and nonsingular covariance matrix κ and, as usual, $S_n^* = n^{-1/2}(Y_1 + \cdots + Y_n)$. Further, $p^{[r]}$ is the direct Edgeworth approximation of order r, defined for a general random vector X by (6.27).

At certain places we impose the assumption that $Y_n = X_n - E_\theta(X_n)$, where $X_1, X_2, \ldots$ are independent and identically distributed, copies of a random variable X, having an exponential model distribution with minimal representation

$$e^{\theta \cdot x - k(\theta) - h(x)}, \tag{6.125}$$

the domain of variation Θ of the canonical parameter θ being an open convex subset of R^m. The following regularity condition will be referred to later.

(c) For every $\theta \in \Theta$ there exists a positive integer n_0 such that $S_n = X_1 + \cdots + X_n$ has a continuous bounded probability density function for $n \geqslant n_0$.

This condition is, by section 6.11, equivalent to

(c′) For every $\theta \in \Theta$ there exists $\lambda \geqslant 1$ such that

$$\int_{R^m} |M(X; i\zeta; \theta)|^\lambda d\zeta < \infty.$$

We first treat the case of direct Edgeworth expansions.

Theorem 6.4

Let $Y_1, Y_2, \ldots$ be independent and identically distributed with mean 0 and covariance matrix κ and suppose that for some positive integer n_0 the standardized sum $S_n^* = n^{-1/2}(Y_1 + \cdots + Y_n)$ has a bounded continuous density $p(s)$ for $n \geqslant n_0$. The latter condition is equivalent to integrability of $|\chi|^\lambda$ for some $\lambda \geqslant 1$, where χ denotes the common characteristic function of the Y_n.

If all moments of Y_n of order $r + 2$ exist, for some integer $r \geqslant 0$, then

$$p(s_n^*) = p^{[r]}(s_n^*) + o(n^{-(1/2)r}) \tag{6.126}$$

uniformly in s_n^*. In case the moments of order $r + 3$ exist the error assessment on the right-hand side of (6.126) may be sharpened to $O(n^{-(r+1)/2})$.

For a proof, see Bhattacharya and Rao (1976, section 19).

By a well known property of exponential families, a random variable X with distribution (6.125) has cumulants of all orders. We denote these by $\kappa_0^{i_1, \dots, i_\nu}(\theta)$, and we write $\kappa_0(\theta)$ for the variance matrix $[\kappa_0^{i,j}(\theta)]$.

Theorem 6.5

Let X_1, $X_2, \dots$ be independent and identically distributed in the exponential model (6.125) and let $S_n^* = n^{-1/2}(X_1 + \dots + X_n - nE_\theta X_1)$. Suppose that condition (c), or equivalently (c'), is satisfied.
Then for any $r = 0, 1, \dots$

$$p(s_n^*; \theta) = p^{[r]}(s_n^*; \theta) + O(n^{-(r+1)/2}) \qquad (6.127)$$

uniformly in s_n^* and in θ on every compact subset of Θ.

We recall that $p^{[r]}(s_n^*; \theta)$, the rth order Edgeworth approximation to $p(s_n^*; \theta)$, is given by

$$p^{[r]}(s_n^*; \theta) = \phi_m(s_n^*; \kappa_0(\theta))\{1 + Q_3(s_n^*; \theta) + \dots + Q_{r+2}(s_n^*; \theta)\},$$

where

$$Q_\nu(s_n^*; \theta) = n^{-\nu/2} Q_{0\nu}(s_n^*; \theta)$$

and where $Q_{0\nu}(\cdot; \theta)$ does not depend on n and is given by

$$Q_{03}(x; \theta) = \tfrac{1}{6}\kappa_0^{i,j,k}(\theta) h_{ijk}(x; \kappa_0(\theta)),$$
$$Q_{04}(x; \theta) = \tfrac{1}{24}\kappa_0^{i,j,k,l}(\theta) h_{ijkl}(x; \kappa_0(\theta)),$$
$$+ \tfrac{1}{72}\kappa_0^{i,j,k}(\theta)\kappa^{l,m,n}(\theta) h_{ijklmn}(x; \kappa_0(\theta))$$

etc.

For an outline of the proof of Theorem 6.5, see Barndorff-Nielsen and Cox (1979, Appendix).

To deal with the case where the Y_n are not necessarily identically distributed, we let

$$\rho_{\nu n} = n^{-1} \sum_{j=1}^{n} E(|Y_j|^\nu),$$

where $|\cdot|$ denotes the Euclidean norm, and with q denoting a positive integer we define functions g_k ($k \geqslant 0$) on R^m by

$$g_k(t) = \prod_{j=k+1}^{k+q} |M(Y_n; it)|.$$

Theorem 6.6

Let λ_n denote the smallest eigenvalue of $\operatorname{cov}(S_n^*)$ and suppose:

(i)
$$\liminf_{n\to\infty} \lambda_n > 0. \tag{6.128}$$

(ii) For some integer $v > 2$
$$\sup_{n\geqslant 1} \rho_{vn} < \infty. \tag{6.129}$$

(iii) For some integer $q > 0$ and for all $c > 0$
$$\sup_{k\geqslant 0} \int g_k(t)dt < \infty, \ \sup\{g_k(t): |t| > c, k \geqslant 0\} < 1. \tag{6.130}$$

Then

$$p(s_n^*) = p^{[r]}(s_n^*) + o(n^{-r/2}) \tag{6.131}$$

uniformly in s_n^*.

For a proof see Bhattacharya and Rao (1976, section 19).

Under the conditions of Theorems 6.4 and 6.6, integrated versions of (6.126) and (6.131) hold as well. Specifically, in the setting of Theorem 6.4 and letting

$$dP_{S_n^*} = p(s_n^*)ds_n^*, \qquad dP_{S_n^*}^{[r]} = p^{[r]}(s_n^*)ds_n^*$$

and noting that $P_{S_n^*}^{[r]}$ is a signed measure, we have

$$\int f\, dP_{S_n^*} = \int f\, dP_{S_n^*}^{[r]} + o(n^{-r/2}), \tag{6.132}$$

for any real-valued function f on R^m such that (i) the derivatives $f^{(I)}$ of f are continuous for $|I| \leqslant r$, (ii) $(1 + |x|^2)^{-r/2+1}|f(x)|$ is bounded away from infinity, (iii) for $|I| = r$ the derivative $f^{(I)}$ has at most polynomial growth at infinity (Bhattacharya, 1985).

Various further results for non-identically distributed and/or dependent base variates have been discussed *inter alia* by Bhattacharya and Rao (1976), Durbin (1980), Götze and Hipp (1983) and Skovgaard (1986).

In particular, to simplify the treatment of multivariate expansions in general and to take account of the possibility that the asymptotic accuracy may vary with the direction in R^m-space, Skovgaard (1986) introduced a directional approach. To describe this and some of the main results of Skovgaard's paper we now consider a sequence Z_1, $Z_2, \ldots$ of m-dimensional random vectors having mean 0 and denote the covariance matrix of Z_n by κ_n. The matrix κ_n is assumed to be nonsingular for every n. For any $t = (t^1, \ldots, t^m) \in R^m$ let

$$\rho_{\nu n}(t) = \sup\left\{\left(\frac{1}{k!}|\chi_{n,k}(t)|/|t|_n^k\right)^{1/(k-2)}; 3 \leqslant k \leqslant \nu\right\}$$

and

$$R_{\nu n}(t) = \sup\left\{\left|\frac{d^{\nu+1}}{dh^{\nu+1}}\log \zeta_n(ht)\right| \bigg/ (|t|_n^{\nu+1}(\nu+1)!); 0 < h < 1\right\},$$

where $|t|_n^2 = t\kappa_n^{-1}t^{\mathrm{T}}$ and where $\chi_{n,k}(t^k)$ is short for

$$\chi_{n,k}(t) = K\{t \cdot Z_n, \ldots, t \cdot Z_n\},$$

where the argument $t \cdot Z_n$ is repeated k times. Note that $\rho_{\nu n}(t)$ depends on t only through its direction $t/|t|_n$. We assume that $\rho_{\nu n}(t) \neq 0$ for all n and t.

Theorem 6.7
Let ν be an integer with $\nu > 2$ and suppose that as $n \to \infty$:

(i) $$\varepsilon_n = \sup\{\rho_{\nu n}(t) : t \in R^m\} \to 0.$$

(ii) For all $\delta > 0$,

$$\int_{|t|_n > \delta/\rho_{\nu n}(t)} |M(Z_n; it)|\, dt = O(\varepsilon_n^{\nu-1}).$$

(iii) For some $\delta > 0$

$$R_{\nu n}(\{\delta/(|t|_n \rho_{\nu n}(t))\}t) = O(\rho_{\nu n}(t)^{\nu-1}),$$

uniformly for $t \in R^m$. Then for all sufficiently large n the variate Z_n possesses a probability density function $p(z_n)$ and

$$p(z_n) = p^{[\nu-2]}(z_n) + O(\varepsilon_n)$$

uniformly for $z_n \in R^m$.

In the case of identical replications, $\rho_{\nu n}(t) = cn^{-(1/2)}$ for some $c > 0$.

Of the conditions (i)–(iii) in Theorem 6.7 it is usually (ii) which is most difficult to check, while (i) is often rather easy to establish. As to (iii), great simplification is achieved if the Laplace transform of Z_n exists in a neighbourhood of the origin. For details and the proof, see Skovgaard (1986).

We now turn to the tilted expansions.

Consider a sequence of exponential models $\mathscr{E}_n$, $n = 1, 2, \ldots,$with model functions

$$p_n(t;\theta) = e^{\theta \cdot t - k_n(\theta) - h_n(t)}, \tag{6.133}$$

where t and θ are vectors of dimension k and the canonical parameter θ ranges over an open and convex subset Θ of R^m, both m and θ being independent of n. Further, for each n, (6.133) is assumed to be a minimal exponential representation of $\mathscr{E}_n$. Let T_n denote a random vector with distribution (6.133). The characteristic function of T_n is $M(T_n; i\zeta; \theta)$. The corresponding maximum likelihood estimator and observed information matrix are denoted by $\hat{\theta}$ and $j(\theta)$, the dependence on n thus being notationally suppressed. Note that $j(\theta) = \nabla\nabla^{\mathrm{T}} k_n(\theta)$.

We refer to the following conditions:

(A) For every $\theta \in \Theta$, $k_n(\theta) = O(n)$.

(B) For every $\theta \in \Theta$ and $\delta > 0$

$$\limsup_{n \geqslant 1} \sup_{|\zeta| \geqslant \delta} |M(T_n; i\zeta; \theta)|^{1/n} < 1.$$

(C) For every $\theta \in \Theta$, there exists a $\lambda > 0$, an $n_0 = n_0(\theta)$ and an ε with $0 < \varepsilon < 1$, such that for $n \geqslant n_0$

$$\int_{R^k} |M(T_n; i\zeta; \theta)|^{\lambda/n} d\zeta = O(e^{n^\varepsilon}).$$

Theorem 6.8

Let T_n $(n = 1, 2, \ldots)$ be a sequence of k-dimensional random vectors such that for some $\theta \in \Theta$ independent of n the distribution of T_n is given by (6.133). Let conditions (A), (B) and (C) be satisfied.

Then for $n \geqslant n_0(\theta)$ the random vector $\bar{T}_n = T_n/n$ has a probability density function $p(\bar{T}_n; t; \theta)$ and

$$p(\bar{T}_n; t; \theta) = (2\pi)^{-k/2} |j(\hat{\theta})|^{-1/2} \times \exp[\{\theta \cdot (nt) - k_n(\theta)\} - \{\hat{\theta} \cdot (nt) - k_n(\hat{\theta})\}]\{1 + O(n^{-1})\}. \tag{6.134}$$

For a proof, see Chaganty and Sethuraman (1986).

Suppose in particular that T_n is the sum of n independent and identically distributed observations from the exponential model (6.125). Then condition (A) is automatically satisfied, and (B) reduces to

$$\sup_{|\zeta|\geq\delta}|M(X;i\zeta;\theta)|<1, \tag{6.135}$$

i.e. Cramér's condition (Cramér, 1937, Chapter 7) for the characteristic function of the random variable X with distribution (6.125). Finally, (C) takes the form

$$\int_{R^m}|M(X;i\zeta;\theta)|^{\lambda}d\zeta=O(e^{n^{\varepsilon}}). \tag{6.136}$$

Condition (c), or the equivalent (c′), stated at the beginning of the present section, entails both (6.135) and (6.136). Further, the conclusion of Theorem 6.8 may be strengthened as follows (cf. Barndorff-Nielsen and Cox, 1979, Appendix).

Theorem 6.9

Let $S_n=X_1+\cdots+X_n$, where $X_1,X_2,\ldots$ is a sequence of independent random variables of dimension m and identically distributed according to (6.125). Suppose that condition (c) or (c′) is satisfied.

Then, for any $r=0,1,\ldots,$

$$p(s_n;\theta)=(2\pi)^{-(1/2)m}[\det\{j(\hat{\theta})\}]^{-1/2}\exp\{\theta\cdot s_n-nk_0(\theta)-\hat{\theta}\cdot s_n+nk_0(\hat{\theta})\}$$
$$\times\{1+Q_4(0;\hat{\theta})+\cdots+Q_{2(r+1)}(0;\hat{\theta})+O(n^{-(r+1)})\} \tag{6.137}$$

uniformly in s_n, provided $\hat{\theta}=\hat{\theta}(s_n)$ belongs to a given, but arbitrary, compact subset of int Θ. Here $Q_{2\nu}(0;\hat{\theta})$ is of order $O(n^{-(\nu-1)})$; more specifically,

$$Q_{2\nu}(0;\theta)=Q_{02\nu}(0;\theta)n^{-(\nu-1)}$$

with $Q_{02\nu}$ defined as in Theorem 6.5. Furthermore,

$$k^*(s_n)=\hat{\theta}\cdot s_n-nk(\theta) \tag{6.138}$$

and

$$\det\{j(\theta)\}=n\det\{\kappa_0(\theta)\}=n\det(\nabla\nabla^{\mathrm{T}}k_0). \tag{6.139}$$

Finally we consider the case of lattice distributed variates, restricting discussion to lattice versions of Theorems 6.5 and 6.9. Suppose the support of the exponential model (6.125) belongs to the m-dimensional lattice Z^m, Z denoting the set of all integers. In place of condition (c) we assume the condition:

(d) the support is not contained in any sublattice of Z^m.

This condition may be shown to be equivalent to

(d′) $M(X; i\zeta; \theta) \neq 1$ for $0 \neq \zeta \in (-\pi, \pi]^m$.

We then have the following results, an outline of their proofs being given by Barndorff-Nielsen and Cox (1979, Appendix).

Theorem 6.10
The conclusions of Theorem 6.5 hold also if the support of the exponential model (6.125) is contained in Z^m provided condition (d) or (d′) is satisfied and provided $p(s_n^*; \theta)$ in formula (6.127) is replaced by $n^{(1/2)m}p(s_n; \theta)$.

Theorem 6.11
The conclusions of Theorem 6.9 hold also if the support of the exponential model (6.125) is contained in Z^m provided condition (d) or (d′) is satisfied.

6.14 Appendix: Legendre transformation

The Legendre transform of a function f on R^m is another function f^* on R^m. The transformed function f^* is convex, and if the original function f is convex and lower semicontinuous then f^{**}, i.e. the Legendre transform of f^*, equals f itself. Thus, in this sense, Legendre transformation is an invertible operation.

Legendre transformed functions occur naturally in the theory of exponential models, particularly in the study of log likelihood functions for such models, in the tilted approximations and in large deviation theory, cf. sections 6.4, 6.5 and 6.6. Among the other subject areas where Legendre transformation is considered are mathematical programming, partial differential equations and variational calculus. In analogy with what is the case for other important transforms, for instance the Laplace transform, the usefulness of the Legendre transformation is to a large extent due to the fact that often an initial problem translates, via the transform, to an equivalent, but more easily handled, problem. For the Legendre transformation two such problems, and also the associated functions f and f^*, are said to be dual. Further, f^* is sometimes called the conjugate of f. It is conceptually helpful to think of the dual or conjugate functions f and f^* as being defined on two different copies of R^m.

Below we list some basic definitions and results concerning the Legendre transform.

Let f denote a function defined on R^m and with values in $(-\infty, +\infty]$, i.e. we allow $+\infty$ as a value. To avoid trivialities we assume that f is not identically $+\infty$. From f we define another function f^*, termed the *Legendre transform* of f and given by

$$f^*(y) = \sup_{x \in R^m} \{x \cdot y - f(x)\} \qquad (y \in R^m). \tag{6.140}$$

Whatever the function f, its Legendre transform f^* is a convex function, i.e.

$$f^*(\lambda y_0 + (1-\lambda)y_1) \leqslant \lambda f^*(y_0) + (1-\lambda) f^*(y_1)$$

for every y_0, $y_1 \in R^m$ and $0 \leqslant \lambda \leqslant 1$.

A convex function on R^m is said to be *closed* if it is lower semi-continuous, i.e. if

$$f(x) \leqslant \liminf_{y \to x} f(y)$$

for every $x \in R^m$. From now on we suppose that f is a closed convex function on R^m. Then the Legendre transform of f^* equals f, in symbols $f^{**} = f$ (Rockafellar, 1970, p. 104).

Let $C = \{x \in R^m : f(x) < \infty\}$ and $D = \operatorname{int} C$, the interior of C. The function f is *strictly convex* on a subset C_0 of C if

$$f(\lambda x_0 + (1-\lambda)x_1) < \lambda f(x_0) + (1-\lambda) f(x_1)$$

for every x_0, $x_1 \in C_0$ and $0 < \lambda < 1$. Finally, we need the concept of steepness. Suppose D is nonempty and that f is differentiable throughout D. Then f is said to be *steep* if as $\lambda \downarrow 0$

$$\frac{d}{d\lambda} f(\lambda x_0 + (1-\lambda)x_1) \to -\infty$$

for every $x_0 \in D$ and every boundary point x_1 of C (Rockafellar, 1970, p. 252).

Theorem 6.12

Let f be a closed convex function on R^m which is strictly convex on C and differentiable on $D = \text{int}\, C$, where $C = \{x: f(x) < \infty\}$, and suppose that f is steep. Let $D^* = \text{int}\, C^*$, where $C^* = \{y: f^*(y) < \infty\}$. The gradient mapping ∇f is then one-to-one from D to D^*, continuous in both directions, and $\nabla f^* = (\nabla f)^{-1}$ (Rockafellar, 1970, p. 258).

This applies in particular to the cumulant transform k of an arbitrary m-dimensional random variable X, and in that case some further strong conclusions can be drawn, cf. Theorem 6.1.

Further results and exercises

6.1 Show that as $n \to \infty$, subject to the convergence of the series on the right-hand side,

$$\int_{-\infty}^{\infty}\int \exp\{n \cos x \cos y\} q(x, y)\, dx\, dy$$

$$= 2\pi n^{-1} e^n \sum_{r,s=-\infty}^{\infty} q(2r\pi, 2s\pi)\{1 + O(n^{-1/2})\}.$$

[Section 6.2]

6.2 Let Z be a random variable with mean 0 and suppose that the cumulants κ_r of Z possess asymptotic expansions

$$\kappa_2 = \kappa_{20} + \kappa_{21} n^{-1/2} + \kappa_{22} n^{-1} + O(n^{-3/2}),$$
$$\kappa_3 = \kappa_{30} n^{-1/2} + \kappa_{31} n^{-1} + O(n^{-3/2}),$$
$$\kappa_4 = \kappa_{40} n^{-1} + O(n^{-3/2}),$$
$$\kappa_r = O(n^{-(r-1)/2}), \qquad r > 4.$$

Find explicit expressions for the first few terms of the Edgeworth and Cornish–Fisher expansions for Z.

[Sections 6.3, 6.10; Withers, 1984; Niki and Konishi, 1986]

6.3 Let $X_1, \ldots, X_n$ be independent random variables, each following an exponential distribution, and suppose the reciprocal of the mean value of X_n is of the form $\alpha + \beta t_n$, where $\alpha > 0$ and $\beta > 0$ are considered as unknown parameters while the t_n are known positive constants (covariates). Then

$$S_n = \left(\sum_{i=1}^{n} X_i, \sum_{i=1}^{n} t_i X_i\right)$$

is the minimal sufficient and canonical statistic. Let m_n and M_n denote the eigenvalues of the variance matrix of S_n, with $m_n \leqslant M_n$. These, of course, depend on α, β and $t_1, \dots, t_n$.

Show that the condition

$$\log n/m_n \to 0$$

as $n \to \infty$ holds either for all (α, β) or for none, and verify that if $t_1, \dots, t_n$ are such that the condition does hold then S_n possesses a valid Edgeworth expansion. Determine the first few terms of this expansion.

Show further that $m_n \to \infty$ as $n \to \infty$ if and only if

$$\sum (t_i - \bar{t})^2 \to \infty \text{ and } \sum t_i^{-2} \to \infty.$$

Discuss the question of an Edgeworth expansion for S_n when the first condition above does not hold but $m_n \to \infty$. What is the situation if $m_n \nrightarrow \infty$?

[Sections 6.3, 6.13; Skovgaard 1981, 1986]

6.4 For a parametric class of probability distributions $\mathscr{P} = \{\mathscr{P}_\theta : \theta \in \Theta\}$, let $I(\theta_0, \theta)$ denote the Kullback–Leibler information of P_θ with respect to P_{θ_0}. Show that under a core exponential model of section 6.4 we have

$$I(\theta_0, \theta) = (\theta_0 - \theta) \cdot \tau_0 - k(\theta_0) + k(\theta),$$
$$I(\hat{\theta}, \theta) = l(\hat{\theta}) - l(\theta), \quad \text{where } l \text{ denotes log likelihood.}$$

[Section 6.4]

6.5 In each of the three types of model below let all the parameters be considered as unknown. Show that for any sample size large enough to ensure existence with probability 1 of the maximum likelihood estimates, the normalized tilted approximation to the distribution of the minimal sufficient and canonical statistic is exact:

(i) the multivariate normal model;

(ii) the bivariate inverse Gaussian model with model function

$$\frac{\sqrt{(\chi\kappa)}}{2\pi} e^{\sqrt{(\chi\psi)} + \sqrt{(\kappa\lambda x)}} (xy^3)^{-1/2} e^{-(1/2)\{\chi x^{-1} + \psi x + \kappa x^2 y^{-1} + \lambda y\}},$$

the parameters, all positive, being χ, ψ, κ and λ;

(iii) the bivariate hyperboloid model with model function

$$\frac{\kappa e^{\kappa}}{2\pi}\sinh x \exp[-\kappa\{\cosh x \cosh \chi - \sinh x \sinh \chi \cos(y-\phi)\}],$$

for $x \geqslant 0, 0 \leqslant y < 2\pi$, the parameters being $\kappa > 0$, $\chi \geqslant 0$ and $\phi \in [0, 2\pi)$.

The last two models possess a number of pleasing and useful distributional properties, on which the exactness proofs may be based; see Jensen (1981), Barndorff-Nielsen (1983, 1988).

[Section 6.5]

6.6 The 'standard model' of collective risk theory describes the total amount of claims in the time interval $(0, t)$ by a compound Poisson process $S(t)$. Such a process may be specified by its cumulant generating function k which is of the form

$$k(\eta) = t\lambda\{e^{h(\eta)} - 1\},$$

where λ is the constant rate of the underlying Poisson process and h is the cumulant generating function of the distribution of the insurance claims. Denote the mean value of this claim distribution by μ. Supposing that premiums flow in continuously at a constant rate r and that the initial capital is u, ruin occurs when $S(T) - rt < u$ for the first time. Let $T(u)$ denote this time and consider the following two measures of risk for the insurance company: $P\{S(t) > s\}$ and $P\{T(u) < t\}$.

Conclude that for $x \geqslant \mu$,

$$P\{S(t) \geqslant t\lambda x\} \leqslant e^{-t\lambda H^*(x)},$$

where H^* is the Legendre transform of $H(\eta) = \exp\{h(\eta)\} - 1$.

Suppose that the equation $\lambda H(\eta) = r\eta$ has a positive root $\eta(r)$ for every $r > \lambda\mu$. Then

$$P\{T(u) \leqslant uy\} \leqslant e^{u\eta(r)} \qquad \text{for } y \geqslant \{\lambda H'(\eta(r)) - r\}^{-1}.$$

In particular, for $y = \infty$ one gets the so-called Lundberg bound for the probability that ruin will occur in finite time.

Furthermore, for large t and $c = \lambda H'(\eta(r))$ one has, on account of the tilted approximation for $S(t)$, the approximation

$$P\{S(t) \geqslant tc + y | S(t) \geqslant tc\} \doteq e^{-\eta(r)y}.$$

The key quantity $\eta(r)$ has an interesting interpretation as an analogue of inverse temperature in statistical mechanics.

[Sections 6.5, 6.6; Martin-Löf, 1986]

6.7 Let X be an m-dimensional random vector with cumulant transform $k(t) = K(X; t)$. Show rigorously that for any (Borel set) $B \subset R^m$

$$P(X \in B) \leqslant \exp\left\{ -\inf_{x \in B} k^*(x) \right\},$$

where $k^*(x) = \sup_{t \in R^m} \{x \cdot t - k(t)\}$ is the Legendre transform of k.

[Section 6.6]

6.8 It is often useful, particularly for more abstract sample spaces, to consider the modification of (6.61) according to which a family of probability measures Q_λ on a sample space $\mathscr{X}$ is said to possess a large deviation property with rate function k^* provided that for every closed subset C of $\mathscr{X}$

$$\liminf_{\lambda \to \infty} \{ -\lambda^{-1} \log Q_\lambda(C) \} \geqslant \inf_{x \in C} k^*(x)$$

and that for every open subset G of $\mathscr{X}$

$$\limsup_{\lambda \to \infty} \{ -\lambda^{-1} \log Q_\lambda(G) \} \leqslant \inf_{x \in G} k^*(x);$$

cf., for instance, Varadhan (1984).

[Section 6.6]

6.9 Let $C_0[0, 1]$ denote the set of continuous functions $x(t)$ on the interval $0 \leqslant t \leqslant 1$ such that $x(0) = 0$, and let P_0 be a probability measure on $C_0[0, 1]$.

In analogy with (6.44) one may attempt to define the exponential tilt of P_0 by

$$\frac{dP_\theta}{dP_0} = e^{\langle \theta, x \rangle - k(\theta)},$$

where $\langle \theta, x \rangle$ is the scalar product

$$\langle \theta, x \rangle = \int_0^t \theta(t) dx(t)$$

and

$$k(\theta) = \log \int e^{\langle \theta, x \rangle} dP_0,$$

θ being thought of as the 'derivative', in some sense, of a function $\zeta \in C_0[0, 1]$, which we indicate as $\theta = \dot{\zeta}$. There are some well-known difficulties in giving a rigorous mathematical meaning to this, but let us proceed heuristically.

Now, let $B(t)$ be standard Brownian motion, let $\varepsilon > 0$, and let P_0 be the probability measure for $X(t) = \varepsilon B(t)$ induced from that of $B(t)$, i.e. from Wiener measure. Then $\langle \theta, X \rangle$ is a normal random variable and, using the fact that $E_0\{dX(s)dX(t)\} = \varepsilon^2 \delta(|t-s|)$, we find

$$k(\theta) = \tfrac{1}{2}\varepsilon^2 \int_0^1 \{\theta(t)\}^2 dt.$$

It is natural to consider

$$k^*(\xi) = \sup_{\theta \in C_0[0,1]} \{\langle \theta, \varepsilon^2 \xi \rangle - k(\theta)\},$$

with $\xi = \dot{\eta}$, $\eta \in C_0[0, 1]$, as a possible definition of the Legendre transform of k. We have

$$\langle \theta, \varepsilon^2 \xi \rangle - k(\theta) = \varepsilon^2 \int_0^1 \theta(t)\{\xi(t) - \tfrac{1}{2}\theta(t)\} dt$$

and brief reflection indicates that the supremum of this quantity is achieved for $\theta = \xi$, i.e.

$$k^*(\xi) = \varepsilon^2 \tfrac{1}{2} \int_0^1 \{\dot{\eta}(t)\}^2 dt.$$

For any $\eta \in C_0[0, 1]$ which is absolutely continuous with $\dot{\eta}$ square integrable, let

$$J(\eta) = \int_0^t \{\dot{\eta}(t)\}^2 dt$$

and let $J(\eta) = \infty$ for any other $\eta \in C_0[0, 1]$.

Comparing this to (6.56) and to Exercise 6.8 suggests that for any closed $C \subset C_0[0, 1]$,

$$\liminf_{\varepsilon \to 0} \{-\varepsilon^2 \log P_0(C)\} \geqslant \inf_{\eta \in C} J(\eta),$$

and for any open set $G \subset C_0[0, 1]$

$$\limsup_{\varepsilon \to 0} \{-\varepsilon^2 \log P_0(G)\} \leqslant \inf_{\eta \in G} J(\eta).$$

This is, in fact, the case. For a proof see, for instance, Varadhan (1984).

[Section 6.6; Schilder, 1966]

6.10 As a univariate example of the generalized formal Edgeworth expansion of (6.87) take the distribution to be expanded as the Weibull distribution with survivor function $\exp(-x^\rho)$ and the base distribution with respect to which the expansion is made as either the unit exponential distribution or the exponential distribution with mean chosen to match exactly that of the Weibull distribution. Use two arguments, first direct expansion of the log density in powers of $(\rho - 1)$ and secondly evaluation of the relevant cumulants followed by application of (6.87).

[Section 6.9]

6.11 Let b_i^j be a nonsingular $m \times m$ matrix with inverse c_j^i, let a_i and a^i be two $m \times 1$ vectors and let $\lambda > 0$ be a constant. Show that for any function f on R^m into $(-\infty, +\infty]$ with Legendre transform f^* we have the following table of Legendre transforms, in which $x = (x^1, \ldots, x^m)$, $y = (y_1, \ldots, y_m)$ and, for instance, $b_i^j x^i$ is short for $(b_i^1 x^i, \ldots, b_i^m x^i)$.

f	f^*
$f(b_i^j x^i)$	$f^*(c_i^j y_j)$
$f(x^i - a^i)$	$f^*(y) + a^i y_i$
$f(x) - a_i x^i$	$f^*(y_i + a_i)$
$\lambda f(x^i)$	$\lambda f^*(y^i/\lambda)$

[Section 6.14]

6.12 Show that the only function on R^m for which $f = f^*$ is $f(x) = \frac{1}{2} x \cdot x$, where f^* is the Legendre transform of f.

[Section 6.14]

Bibliographic notes

The multivariate version of Laplace's method is discussed in Bleistein and Handelsman (1975).

For a discussion of the multivariate version of the method of stationary phase (cf. Further results and exercises 3.4) and some of its applications see Guillemin and Sternberg (1977).

The literature contains many papers discussing the asymptotic validity of the Edgeworth approximations under various conditions. Skovgaard (1986) provides a very general treatment. For the case of sums of independent random vectors a comprehensive account of results and of the historical development has been given by Bhattacharya and Rao (1976). Feller (1966, Ch. 16) contains a very lucid treatment of Edgeworth expansions for sequences of one-dimensional, independent and identically distributed variates.

The idea of exponential tilting as a tool for obtaining approximations (section 6.5) is due to Esscher (1932); see also Esscher (1963). Further interesting applications of exponential tilting, in particular to collective risk theory and to queueing theory, may be found in Asmussen (1982, 1984).

The results that underlie tilted asymptotic expansions for densities are local limit results in the sense of the theory of large deviations; see Theorems 6.8, 6.9 and 6.11. This theory, which was initiated by Khinchin (1929) and Cramér (1937), has been undergoing strong developments during the last decade in connexion with applications to statistical mechanics, diffusion processes, random fields, dynamical systems, risk theory, and statistics. See, for instance, Varadhan (1984), Stroock (1984), Ellis (1985), Föllmer (1987), Freidlin and Wentzell (1984), Martin-Löf (1986), Bahadur (1971), and Groeneboom and Oosterhoff (1981).

Volterra seems to have been the first (in 1887) to have considered differentiation of functionals, as in section 6.8; see Volterra (1931).

The material in section 6.9 is developed from McCullagh (1987, Section 5.2).

Section 6.10 is based on Barndorff-Nielsen and Blæsild (1988b).

For further information on the Legendre transform and on its use in the theory of exponential models, see Rockafellar (1970) and Barndorff-Nielsen (1978), Barndorff-Nielsen and Blæsild (1983).

CHAPTER 7

Expansions for conditional distributions

7.1 Introduction

Quite often, especially in applications to statistical inference, the probability distribution of interest is a conditional one. It may be possible to study this directly, for example via explicit expressions for the conditional density, the conditional cumulant generating function or via conditional cumulants. More commonly, however, one has to study the conditional distribution via its expression as a ratio of an overall joint density to the marginal density of the conditioning statistic.

When both densities have to be approximated it is not necessary to use the same type of approximation for the two parts. We shall consider three possibilities: Edgeworth/Edgeworth; partially tilted/direct Edgeworth; tilted/tilted. Here the second, for example, involves applying a mixed expansion, in the sense of section 6.7, to the joint distribution in the numerator and a direct Edgeworth expansion to the denominator. This may be the most effective approach when the marginal distribution of the conditioning statistic is not available in directly accessible form.

We denote by $X_n = (X_{n1}, X_{n2})$ the random variable of interest, partitioned into two components of dimensions p and q, respectively. We suppose that $E(X_n) = 0$. Often, although not necessarily, X_n will be a standardized sum of independent random vectors. For notational simplicity we shall often omit the index n, for instance writing $X = (X_1, X_2)$ rather than $X_n = (X_{n1}, X_{n2})$. We let indices $\alpha, \beta, \ldots$, $a, b, \ldots$ and $r, s, \ldots$ refer to components of X, X_1 and X_2, respectively. In particular, then, $\kappa^{\alpha,\beta}$, $\kappa^{a,b}$ and $\kappa^{r,s}$ indicate respectively the covariance matrices of X, X_1 and X_2.

We wish to approximate

$$f_{X_2|X_1}(x_2|x_1) = f_{X_1,X_2}(x_1,x_2)/f_{X_1}(x_1). \tag{7.1}$$

Now the linear least squares regression coefficients of X_2 on X_1 are $\beta^r_a = \kappa^{r,b}\kappa_{a,b}$, where the $\kappa^{r,b}$ specify the covariances of the components of X_2 and X_1 and $\kappa_{a,b}$ is the $p \times p$ inverse to the covariance matrix of X_1. Thus the residual of X_2 from linear least squares regression on X_1 is

$$\bar{X}^r_2 = X^r_2 - \beta^r_a X^a_1. \tag{7.2}$$

The covariance matrix of this residual is

$$\bar{\kappa}^{r,s} = \kappa^{r,s} - \kappa^{r,a}\kappa^{s,b}\kappa_{a,b}. \tag{7.3}$$

In the simple case when both X_1 and X_2 are scalar, (7.3) reduces to

$$\text{var}(\bar{X}_2) = \text{var}(X_2) - \{\text{cov}(X_1, X_2)\}^2/\text{var}(X_1). \tag{7.4}$$

7.2 Direct–direct expansions

We now apply direct Edgeworth expansions separately to numerator and denominator of (7.1). Let $\kappa = [\kappa^{\alpha,\beta}]$, $\kappa_{(1)} = [\kappa^{a,b}]$ and $\kappa_{(2)} = [\kappa^{r,s}]$, and denote the covariance matrix of the linear regression residual $\bar{X}_2$ by $\bar{\kappa}_{(2)} = [\bar{\kappa}^{r,s}]$. Assuming that the X's have been standardized to make the covariance matrices $O(1)$, it can be shown that

$$f_{X_2|X_1}(x_2|x_1) \doteq \phi_q(\bar{x}_2; \bar{\kappa}_{(2)})\{1 + Q_3(x_2|x_1) + Q_4(x_2|x_1)\}, \tag{7.5}$$

where

$$Q_3(x_2|x_1) = \tfrac{1}{6}\{\kappa^{\alpha,\beta,\gamma}h_{\alpha\beta\gamma}(x;\kappa) - \kappa^{a,b,c}h_{abc}(x_1;\kappa_{(1)})\}, \tag{7.6}$$

$$\begin{aligned} Q_4(x_2|x_1) = \tfrac{1}{72}[3\{&\kappa^{\alpha,\beta,\gamma,\delta}h_{\alpha\beta\gamma\delta}(x;\kappa) - \kappa^{a,b,c,d}h_{abcd}(x_1;\kappa_{(1)})\} \\ &+ \kappa^{\alpha,\beta,\gamma}\kappa^{\delta,\varepsilon,\eta}h_{\alpha\beta\gamma\delta\varepsilon\eta}(x;\kappa) - \kappa^{a,b,c}\kappa^{d,e,f}h_{abcdef}(x_1;\kappa_{(1)}) \\ &- 2\kappa^{a,b,c}h_{abc}(x_1;\kappa_{(1)})\{\kappa^{\alpha,\beta,\gamma}h_{\alpha\beta\gamma}(x;\kappa) \\ &- \kappa^{a,b,c}h_{abc}(x_1;\kappa_{(1)})\}], \end{aligned} \tag{7.7}$$

where under repeated sampling these are respectively of orders $n^{1/2}$ and n^{-1}.

Note particularly the presence of tensorial Hermite polynomials of two different dimensions and thus associated with two different covariance matrices.

It is, of course, possible to develop the expansion of the conditional density to higher orders than in (7.7). To ensure the validity of the conditional expansion which includes the term in $n^{-(r-1)/2}$ and thus has additive error $O(n^{-r/2})$ we have to require that the joint density of X_1 and X_2 has an Edgeworth expansion that is 'twice as long', i.e. which includes the n^{-r+1} term, even though only lower order terms up to $n^{-(r-1)/2}$ enter the conditional expansion; see, for instance, the above formulae for $Q_3(x_2|x_1)$ and $Q_4(x_2|x_1)$. The reason for this rather strong requirement lies in the relative rather than absolute nature of the error terms in the Edgeworth expansions. For more detailed discussion, see Michel (1979) and Hipp (1984).

The simplest special case of these formulae arises when $p = q = 1$. There is no essential loss of generality in supposing X_1 and X_2 to have unit variance, so that in particular the tensorial Hermite polynomials associated with $\kappa_{(1)}$ are ordinary Hermite polynomials. Then, examining for simplicity just the $n^{-1/2}$ term, we have that

$$Q_3(x_2|x_1) = \tfrac{1}{6}\{\kappa^{1,1,1}h_{111}(x;\kappa) + 3\kappa^{1,1,2}h_{112}(x;\kappa) + 3\kappa^{1,2,2}h_{122}(x;\kappa) + \kappa^{2,2,2}h_{222}(x;\kappa) - \kappa^{1,1,1}H_3(x_1)\}; \tag{7.8}$$

note the inevitably strong connexion with (6.25).

It follows from (6.26) that in the further special case when $\kappa^{1,2} = 0$, so that X_1 and X_2 are uncorrelated as well as having unit variance and $\bar{X}_2 = X_2$, (7.8) becomes, now in terms of standardized cumulants,

$$Q_3(x_2|x_1) = \tfrac{1}{6}\{3\rho^{1,1,2}H_2(x_1)H_1(x_2) + 3\rho^{1,2,2}H_1(x_1)H_2(x_2) + \rho^{2,2,2}H_3(x_2)\}. \tag{7.9}$$

In this special case the conditional mean and variance of the approximating Edgeworth expansion are respectively

$$\tfrac{1}{2}\rho^{1,1,2}H_2(x_1)/\sqrt{n} = \tfrac{1}{2}\rho^{1,1,2}(x_1^2 - 1)/\sqrt{n},$$
$$1 + \rho^{1,2,2}H_1(x_1)/\sqrt{n} = 1 + \rho^{1,2,2}x_1/\sqrt{n}. \tag{7.10}$$

The form of the mean reinforces the interpretation of $\rho^{1,1,2}$ mentioned in section 5.2 as concerned with the quadratic regression of X_2 on X_1, whereas the form of the variance relates the mixed cumulant, $\rho^{1,2,2}$ to the change with x_1 in the conditional variance of X_2 given $X_1 = x_1$. Of course the same interpretations apply with the roles of X_1 and X_2 interchanged, so that both mixed third-order cumulants have two interpretations in this special case.

7.3 Expansions for conditional cumulants

In section 7.2 we obtained in a very special case the conditional mean and variance of X_2 given $X_1 = x_1$. Now essentially the same calculation can be carried out in the quite general case and moreover for cumulants of higher order. For simplicity we shall here assume that X_1 and X_2 are uncorrelated, i.e. the results can be regarded as applying to X_1 and $\bar{X}_2$. The calculations are tedious and we shall just record the results. For details, see McCullagh (1987, Ch. 5).

We write $\kappa_{2\cdot 1}^{r,s,\ldots}$ for the conditional cumulants. Then with error $O(n^{-3/2})$ we have the following results for cumulants up to order four, terms of higher order being zero with error $O(n^{-3/2})$:

$$\kappa_{2\cdot 1}^{r} = \tfrac{1}{2}\kappa^{r,a,b}h_{ab} + \tfrac{1}{12}\{\kappa^{r,a,b}\kappa^{c,d,e}(h_{abcde} - h_{ab}h_{cde}) + 2\kappa^{r,a,b,c}h_{abc}\}, \tag{7.11}$$

$$\kappa_{2\cdot 1}^{r,s} = \kappa^{r,s} + \kappa^{r,s,a}h_a + \tfrac{1}{12}\{6\kappa^{r,s,a,b}h_{ab} + 2\kappa^{r,s,a}\kappa^{b,c,d}(h_{abcd} - h_a h_{bcd}) + 3\kappa^{r,a,b}\kappa^{s,c,d}(h_{abcd} - h_{ab}h_{cd})\}, \tag{7.12}$$

$$\kappa_{2\cdot 1}^{r,s,t} = \kappa^{r,s,t} + \tfrac{1}{2}\{2\kappa^{r,s,t,a}h_a + \kappa^{r,s,a}\kappa^{t,b,c}(h_{abc} - h_a h_{bc})[3]\}, \tag{7.13}$$

$$\kappa_{2\cdot 1}^{r,s,t,u} = \kappa^{r,s,t,u} + \kappa^{r,s,a}\kappa^{t,u,b}(h_{ab} - h_a h_b)[3]. \tag{7.14}$$

Throughout, the tensorial Hermite polynomials have argument x_1 and are calculated for the covariance matrix $\kappa_{(1)}$. Note that (7.11) gives the conditional mean of X_2 given $X_1 = x_1$, i.e. the regression of X_2 on X_1, the linear component being absent by assumption. Thus the $n^{-1/2}$ term is quadratic in x_1 with coefficients determined by the third-order cumulants, generalizing the first interpretation in (7.10); the polynomial $h_{abcde} - h_{ab}h_{cde}$ defining the n^{-1} term is cubic in x_1. For the conditional variance the $n^{-1/2}$ term is linear in x_1 with coefficients determined by third-order cumulants, generalizing the second interpretation in (7.10). The n^{-1} term in the conditional variance is quadratic, the n^{-1} term in the conditional skewness is linear, and that in the conditional kurtosis constant.

We assumed above, for the formulae (7.11)–(7.14), that X_1 and X_2 are uncorrelated. If they are not, one may, as already noted, apply (7.11)–(7.14) to X_1 and $\bar{X}_2$, where $\bar{X}_2^r$, given by (7.2), is the residual of X_2 after regression on X_1. For this and other purposes it is useful to have a list of the relevant cumulants of $\bar{X}_2$ in terms of those of X_1 and X_2, and we provide the formulae in question below. Again, we refer to McCullagh (1987, Ch. 5) for a detailed derivation:

$$\bar{\kappa}^{r,s} = \kappa^{r,s} - \beta^r_a \beta^s_b \kappa^{a,b}, \qquad \bar{\kappa}^{a,r} = 0, \tag{7.15}$$

$$\begin{aligned}
\bar{\kappa}^{r,s,t} &= \kappa^{r,s,t} - \beta^r_a \kappa^{a,s,t}[3] + \beta^r_a \beta^s_b \kappa^{a,b,t}[3] - \beta^r_a \beta^s_b \beta^t_c \kappa^{a,b,c},\\
\bar{\kappa}^{a,r,s} &= \kappa^{a,r,s} - \beta^r_b \kappa^{a,b,s}[2] + \beta^r_b \beta^s_c \kappa^{a,b,c},\\
\bar{\kappa}^{a,b,r} &= \kappa^{a,b,r} - \beta^r_c \kappa^{a,b,c},
\end{aligned} \tag{7.16}$$

$$\begin{aligned}
\bar{\kappa}^{r,s,t,u} &= \kappa^{r,s,t,u} - \beta^r_a \kappa^{a,s,t,u}[4] + \beta^r_a \beta^s_b \kappa^{a,b,t,u}[6]\\
&\quad - \beta^r_a \beta^s_b \beta^t_c \kappa^{a,b,c,u}[4] + \beta^r_a \beta^s_b \beta^t_c \beta^u_d \kappa^{a,b,c,d},\\
\bar{\kappa}^{a,r,s,t} &= \kappa^{a,r,s,t} - \beta^r_b \kappa^{a,b,s,t}[3] + \beta^r_b \beta^s_c \kappa^{a,b,c,t}[3] - \beta^r_b \beta^s_c \beta^t_d \kappa^{a,b,c,d},\\
\bar{\kappa}^{a,b,r,s} &= \kappa^{a,b,r,s} - \beta^r_c \kappa^{a,b,c,s}[2] + \beta^r_c \beta^s_d \kappa^{a,b,c,d},\\
\bar{\kappa}^{a,b,c,r} &= \kappa^{a,b,c,r} - \beta^r_d \kappa^{a,b,c,d}.
\end{aligned} \tag{7.17}$$

Example 7.1 Conditional distributions in the bivariate log normal distribution

A convenient example for which we can compare exact and approximate cumulants is provided by the bivariate log normal distribution; see Examples 5.1, 5.6 and 6.2. In particular, (5.8) gives the conditional mean and variance of Y^2 given $Y^1 = y^1$, showing a power-law dependence on y^1.

As the simplest illustration of the relation between exact and approximate results we examine the conditional mean (7.11) taking only the leading and $n^{-1/2}$ terms. The Hermite polynomial involved is univariate with argument $y^1 - \kappa^1$ and in fact

$$h_1 = (y^1 - \kappa^1)/\kappa^{1,1}, \qquad h_{11} = \{(y^1 - \kappa^1)/\kappa^{1,1}\}^2 - 1/\kappa^{1,1}.$$

The random variable $\bar{Y}_2$ is here

$$(Y^2 - \kappa^2) - \kappa^{1,2}(Y^1 - \kappa^1)/\kappa^{1,1}.$$

For $r = a = b = 1$, a direct calculation via (5.11), or use of (7.16), shows that

$$\bar{\kappa}^{r,a,b} = \exp(2\mu_1 + \mu_2 + 2\sigma_{20} + \tfrac{1}{2}\sigma_{02})(e^{\sigma_{11}} - 1)(e^{\sigma_{11}} - e^{\sigma_{20}}),$$

where the μ_i and σ_{ij} refer to the bivariate normal distribution that generates the bivariate log normal distribution. Thus the leading and first correction terms give for $E(Y^2|Y^1=y^1)$

$$e^{\mu_2+(1/2)\sigma_{20}}+\{e^{\mu_2-\mu_1+(1/2)\sigma_{02}-(1/2)\sigma_{20}}(e^{\sigma_{11}}-1)/(e^{\sigma_{20}}-1)\}$$
$$\times(y^1-e^{\mu_1+(1/2)\sigma_{20}})$$
$$+\bar{\kappa}^{r,a,b}\left[\left\{\frac{y^1-e^{\mu_1+(1/2)\sigma_{20}}}{e^{2\mu_1+\sigma_{20}}(e^{\sigma_{20}}-1)}\right\}^2-e^{-2\mu_1-\sigma_{20}}(e^{\sigma_{20}}-1)^{-1}\right].$$

Now the exact result is given by (5.8) and the present result gives the corresponding Taylor expansion around $y^1=\exp(\mu_1+\frac{1}{2}\sigma_{20})$. As a qualitative check note that for $\sigma_{11}>0$, $\bar{\kappa}^{r,a,b}>0$ if and only if $\sigma_{11}>\sigma_{20}$ and this from (5.8) is the condition for the regression to be convex.

7.4 Tilted–tilted expansions

In section 7.2 we applied separate direct Edgeworth expansions to the two factors determining the conditional density, thus obtaining an expansion in powers of $n^{-1/2}$. If it is feasible to apply separate tilted expansions to the two factors an expansion in powers of n^{-1} will result and in particular the leading term will have an error $O(n^{-1})$ rather than the error $O(n^{-1/2})$ of the normal approximation. To do this the cumulant generating functions and associated maximum likelihood estimators must be available, at least to sufficient approximation, both for (X_1, X_2) and for X_1 marginally.

The resulting approximation for the conditional density of X_2 given $X_1=x_1$ is thus, from (6.50) and (6.46)–(6.48),

$$\frac{\exp\{k_n(\hat{\lambda})-k_n(\hat{\lambda}_{10},0)-\hat{\lambda}\cdot x+\hat{\lambda}_{10}\cdot x_1\}}{(2\pi)^{q/2}[\det\{j_n(\hat{\lambda})\}/\det\{j_{n1}(\hat{\lambda}_{10})\}]^{1/2}}$$
$$\times\left[1+\frac{1}{n}\{Q_4(0;\hat{\kappa}^{(4)})-Q_4(0;\hat{\kappa}^{(4)}_{10})\}+O(n^{-2})\right]. \qquad (7.18)$$

Here $j_n(\hat{\lambda})$ is the observed $n\times n$ information matrix in the full exponential tilt, whereas $j_{n1}(\hat{\lambda}_{10})$ is the corresponding $p\times p$ matrix in the partial exponential tilt referring to X_1 marginally.

When the dimension q of X_2 is 1 an approximation to the conditional distribution function of X_2 can be derived by the same technique as was used for formulae (4.71) and (4.72). For continuous

X one obtains, with error $O(n^{-1})$,

$$P(X_2 \leqslant x_2 | X_1 = x_1) \doteq \Phi(r) - (r^{-1} - v^{-1})\phi(r), \qquad (7.19)$$

where

$$r = \operatorname{sgn} \hat{\lambda}_2 \sqrt{n}[2\{\hat{\lambda}\cdot x - \hat{\lambda}_{10}\cdot x_1 - k_n(\hat{\lambda}) + k_n(\hat{\lambda}_{10}, 0)\}]^{1/2}$$

and

$$v = \hat{\lambda}_2 / [\det\{j_{n1}(\hat{\lambda}_{10})\}/\det\{j_n(\hat{\lambda})\}]^{1/2},$$

i.e. r and v are, respectively, the signed likelihood ratio statistic and a score statistic for testing $\lambda_2 = 0$. Note that (7.19) is formally identical to (4.71) to which it reduces if X_1 and X_2 are independent, since exponential tilting preserves the independence of X_1 and X_2. For proof and details, see Skovgaard (1987).

7.5 Mixed expansions

It may sometimes be impracticable to work with the part of the tilted calculations that refers to X_2 and then it is likely to be advantageous to use the mixed direct–tilted expression of section 6.7 for the joint distribution of (X_1, X_2) combined with the full tilted expansion for the marginal distribution of X_1. There results the tilted–Edgeworth/tilted expansion

$$\phi_q\{x_2 - \mu_2(\hat{\lambda}_{10}, 0); \kappa_{2\cdot 1}(\hat{\lambda}_{10}, 0)\}$$
$$\times \left[1 + \frac{Q_3\{0, x_2 - \mu_2(\hat{\lambda}_{10}, 0); \kappa^{(3)}(\hat{\lambda}_{10}, 0)\}}{\sqrt{n}}\right.$$
$$\left. + \frac{Q_4\{0, x_2 - \mu_2(\hat{\lambda}_{10}, 0); \kappa^{(4)}(\hat{\lambda}_{10}, 0)\} - Q_4\{0; \kappa^{(4)}_{10}(\hat{\lambda}_{10}, 0)\}}{n}\right]$$
$$+ O(n^{-3/2}), \qquad (7.20)$$

the leading term of which is a normal distribution. Here $\mu_2(\hat{\lambda}_{10}, 0)$ and $\kappa_{2\cdot 1}(\hat{\lambda}_{10}, 0)$ are the unconditional mean of X_2 and the conditional variance of X_2 given $X_1 = x_1$ evaluated for the exponential tilt $(\hat{\lambda}_{10}, 0)$; also $\kappa^{(r)}(\hat{\lambda}_{10}, 0)$ and $\kappa^{(r)}_{10}(\hat{\lambda}_{10}, 0)$ denote the cumulants up to order r of $X = (X_1, X_2)$ and X_1 under that same distribution.

Example 7.2 Time-dependent Poisson process

A Poisson process with rate function $\rho(t)$ is observed over a fixed time interval $[0, t_0]$ and a total of n events are observed, at the time points $t_1, \ldots, t_n$. Suppose that to check the adequacy of the log-linear specification $\rho(t) = \exp(\alpha + \beta t)$ it is desired to test $\gamma = 0$ under the extended hypothesis

$$\rho(t) = \exp(\alpha + \beta t + \gamma t^2).$$

It is natural to carry out the test in the conditional distribution of $\sum t_i^2$ given n and $\sum t_i$. Now, conditioning first on n reduces the problem to that of testing $\gamma = 0$ conditional on $\sum t_i$ under the exponential model

$$n! \exp\{\beta \sum t_i + \gamma \sum t_i^2 - n k_0(\beta, \gamma)\}, \tag{7.21}$$

where

$$k_0(\beta, \gamma) = \log \int_0^{t_0} e^{\beta t + \gamma t^2} dt. \tag{7.22}$$

To determine the leading, normal term of the mixed tilted–Edgeworth approximation (7.20) to the conditional distribution of $\sum t_i^2$ given $\sum t_i$ we need to find the mean and covariance matrix of $(\sum t_i, \sum t_i^2)$ under (7.21), with $\gamma = 0$, and expressions for these are readily obtained by differentiation of the cumulant function (7.22).

For some numerical examples see Pedersen (1979).

Example 7.3 von Mises distribution

As another instance of the derivation of a test by extension of the exponential family, consider the testing of consistency with the von Mises distribution written in canonical form with density for the random angle Y proportional to

$$\exp(\lambda_1 \cos y + \lambda_2 \sin y) = \exp\{\kappa \cos(y - \phi)\}, \tag{7.23}$$

say. One natural extension is to augment (7.23) by terms in $\cos(2y)$ and $\sin(2y)$, i.e. to consider the density proportional to

$$\exp\{\lambda_1 \cos y + \lambda_2 \sin y + \psi_1 \cos(2y) + \psi_2 \sin(2y)\}. \tag{7.24}$$

Unfortunately the normalizing constant and hence the cumulant generating function cannot be written in useful explicit form.

To examine the hypothesis $\psi_1 = \psi_2 = 0$ on the basis of n independent observations, we consider the conditional distribution of $V = (\sum \cos(2Y_i), \sum \sin(2Y_i))$ given $U = (\sum \cos Y_i, \sum \sin Y_i) = u$. The leading term of the mixed approximation (7.20) is that V is bivariate normal.

It is simplest to begin by calculating the covariance matrix of $(\cos Y, \cos(2Y), \sin Y, \sin(2Y))$ from (7.23) with $\phi = 0$. On introducing the Bessel functions

$$I_m(\kappa) = \frac{1}{2\pi}\int_{-\pi}^{\pi} \cos(mx) e^{\kappa \cos x} dx, \tag{7.25}$$

we have that the sine terms are uncorrelated with the cosine terms and that conditional variances from the least squares regression of $\cos(2Y)$ on $\cos Y$ and of $\sin(2Y)$ on $\sin Y$ are respectively

$$\begin{aligned} v_{2\cdot 1}^{c}(\kappa) &= \frac{I_0^2 + I_0 I_4 - 2I_2^2}{2I_0^2} - \frac{(I_0 I_3 + I_0 I_1 - 2I_1 I_2)^2}{2I_0^2(I_0^2 + I_0 I_2 - 2I_1^2)}, \\ v_{2\cdot 1}^{s}(\kappa) &= \frac{(I_0 - I_4)(I_0 - I_2) - (I_1 - I_3)^2}{2I_0(I_0 - I_2)}, \end{aligned} \tag{7.26}$$

where the Bessel functions have argument κ.

Now the maximum likelihood estimates of (λ_1, λ_2), or equivalently (κ, ϕ), in (7.23) satisfy

$$n^{-1}\sum \sin(y_i - \hat{\phi}) = 0, \qquad n^{-1}\sum \cos(y_i - \hat{\phi}) = \hat{I}_1/\hat{I}_0,$$

where the Bessel functions now have argument $\hat{\kappa}$. It follows that the leading term in the approximation to the required conditional distribution is such that

$$\sum \cos\{2(Y_i - \hat{\phi})\} - n\hat{I}_1/\hat{I}_0, \qquad \sum \sin\{2(Y_i - \hat{\phi})\} \tag{7.27}$$

are independently normally distributed with variances $nv_{2\cdot 1}^{c}(\hat{\kappa})$ and $nv_{2\cdot 1}^{s}(\hat{\kappa})$. A chi-squared statistic with two degrees of freedom can thus be formed.

Further results and exercises

7.1 Let X and Y be random variables with joint characteristic function $\chi(X, Y; s, t)$. The characteristic function for Y under the conditional distribution for Y given $X = x$ may if X is discrete, taking integer values only, be expressed as

$$\chi(Y;t|x)=\frac{\int_{-\pi}^{\pi}\chi(X,Y;s,t)e^{-isx}ds}{\int_{-\pi}^{\pi}\chi(X,Y;s,0)e^{-isx}ds}, \tag{7.28}$$

while if X is continuous

$$\chi(Y;t|x)=\frac{\int_{-\infty}^{+\infty}\chi(X,Y;s,t)e^{-isx}ds}{\int_{-\infty}^{+\infty}\chi(X,Y;s,0)e^{-isx}ds} \tag{7.29}$$

(Bartlett, 1938). These formulae are the starting point of much work on conditional asymptotic behaviour; see, for instance, Holst (1979, 1981), Kudlaev (1984) and also Jensen (1987).

[Section 7.1]

7.2 Let $(X_1, Y_1),\ldots,(X_n, Y_n)$ be independent identically distributed copies of a random variate (X, Y) of dimension 2, and suppose that X takes nonnegative integer values only and that X and Y have finite second moments and $E(Y)=0$.

Show that

$$E(Y_+|x_+)=n\{2\pi P(X_+=x_+)\}^{-1}\int_{-\pi}^{\pi}E(Ye^{isX})\chi(X;s)^{n-1}e^{-isx_+}ds$$

and

$$\begin{aligned}E(\textstyle\sum Y_j^2|x_+)&=n\{2\pi P(X_+=x_+)\}^{-1}\\&\quad\times\int_{-\pi}^{\pi}[E(Y^2e^{isX})\chi(x;s)+(n-1)E(Ye^{isX})^2]\\&\quad\times\{\chi(X;s)\}^{n-2}e^{-isx_+}ds\end{aligned}$$

(cf., for instance, Swensen, 1983).

[Section 7.1]

7.3 Let Y_1, Y_2, Y_3 have independent Poisson distributions with means $n\mu_1$, $n\mu_2$, $n\mu_3$. Let $S=Y_1+Y_2+Y_3$, $S_{13}=Y_1+Y_3$, $S_{23}=Y_2+Y_3$. Examine the conditional distributions of Y_3 given (a) $S=s$, (b) $S_{13}=s_{13}$ and $S_{23}=s_{23}$. Compare the exact distributions with various approximations derived via direct Edgeworth and tilted methods, applying as $n \to \infty$.

[Sections 7.2, 7.4, 7.5]

7.4 Let $Y_1, \ldots, Y_n$ be independent binary random variables with

$$P(Y_j = 0) = (1 + e^{\alpha + \beta x_j})^{-1}, \qquad P(Y_j = 1) = e^{\alpha + \beta x_j}(1 + e^{\alpha + \beta x_j})^{-1},$$

where $x_1, \ldots, x_n$ are given constants. Show that the exact conditional distribution of $\sum x_j Y_j$ given $\sum Y_j = s$ has the form

$$P(\sum x_j Y_j = z | \sum Y_j = s) = e^{z\beta} c(s, z) / \sum_u e^{u\beta} c(s, u)$$

where $c(s, u)$ is the number of distinct choices of s values from the finite population $\{x_1, \ldots, x_n\}$ such that the sum of the x's is u. Examine double Edgeworth and double tilted approximations to the conditional distribution.

[Sections 7.2, 7.4; Daniels, 1958]

7.5 Consider again the procedure of testing the von Mises model by exponential extension as in Example 7.3. As was noted there, the norming constant associated with (7.24) is not available in any explicit form. In particular, therefore, the double tilted method of section 7.4 cannot be immediately applied. However, if in (7.25) the value of $\kappa = \sqrt{(\lambda_1^2 + \lambda_2^2)}$ is large compared to ψ_1 and ψ_2 then Laplace's method may be used to obtain an approximation to the norming constant. Compare the resulting approximation of the conditional distribution of $(\sum \cos(2A_i), \sum \sin(2A_i))$ to the normal approximation derived in Example 7.3.

[Section 7.5]

Bibliographic notes

Exact conditions for the validity of asymptotic expansions for conditional distributions are given in Michel (1979), Hipp (1984), Landers and Rogge (1986) and Skovgaard (1987). Michel (1979) and Hipp (1984) discuss the case of Edgeworth/Edgeworth expansions for sums of independent identically distributed vectors, while Landers and Rogge (1986) consider more general conditioning events.

The initial question of determining the conditional limit distribution or, in other words, the leading asymptotic term, has been studied by Steck (1957), Holst (1979, 1981), Swensen (1983), and Kudlaev (1984); see also Landers and Rogge (1987) and references given there.

Postscript

There are two complementary themes implicit in this book. The first is that for relatively simple problems, especially univariate problems, there are powerful simply applied techniques for finding good approximations to a range of problems in statistical theory and method and in applied probability. Moreover, especially with judicious use of the device of amalgamation into the leading terms and of tilted (saddlepoint) expansions, approximations can be obtained that are frequently appreciably better than those via the crudest use of linearization and asymptotic normality. Sometimes, even, an initial approximate analysis may suggest the form of an 'exact' solution to the problem under study.

The second theme, predominant in the later chapters of the book, is that the apparently much more formidable analogous multivariate problems can, via a suitable notation, be shown to have solutions formally virtually identical to the corresponding univariate ones. This is both of considerable conceptual importance and also of value in applications in its potential for simplifying some of the very lengthy higher-order calculations. This is likely ultimately to be achieved by the incorporation of the techniques into a suitable system of computerized algebra; it has to be admitted that in many problems of realistic complexity the direct translation of the simple general formulae into specific form is a major task.

Our objective throughout is the provision of good approximate solutions to complicated problems. In applications some judgement is usually required, the sharp bounds on error that would make assessment of adequacy quite objective being rarely available. That is, supplementation by qualitative considerations, by numerical work and by comparison with experience with similar problems is often needed. In particular, this implies that, without substantial extra evidence, conclusions drawn from very refined comparisons of rates of convergence may not be a sound guide.

Applications, realized and potential, of these ideas fall into three main groups. Specific investigations in applied statistics often call for relatively minor adaptations of general ideas and easily implemented efficient techniques of approximation can be of immediate value. Secondly, many of the more difficult problems in statistical theory call for higher-order calculations, both to provide a basis of choice between procedures equivalent to the first order of asymptotic theory and to refine the resulting distribution theory. Finally, the study of specific problems in applied probability often benefits from asymptotic analysis of the resulting systems. In a subsequent volume we hope to return especially to the second of these groups of application where the full generality of the methods discussed here is necessary.

References

Abramowitz, M. and Stegun, I.A. (1965) *Handbook of Mathematical Functions*. Dover, New York.

Aigner, M. (1979) *Combinatorial Theory*. Springer, Heidelberg.

Amari, S.-I. and Kumon, M. (1983) Differential geometry of Edgeworth expansion in curved exponential family. *Ann. Inst. Statist. Math.*, **35**, 1–24.

Anscombe, F.J. (1948) The transformation of Poisson, binomial and negative-binomial data. *Biometrika*, **35**, 246–254.

Asmussen, S. (1982) Conditional limit theorems relating a random walk to its associate, with applications to risk reserve processes and the GI/G/1 queue. *Adv. Appl. Prob.*, **14**, 143–170.

Asmussen, S. (1984) Approximations for the probability of ruin within finite time. *Scand. Actuarial J.*, 31–57.

Avram, F. and Taqqu, M.S. (1987) Noncentral limit theorems and Appell polynomials. *Ann. Probab.*, **15**, 767–775.

Baddeley, A. and Møller, J. (1987) Nearest-neighbour Markov point processes and random sets. To appear in *Int. Statist. Review*.

Bahadur, R.R. (1971) *Some Limit Theorems in Statistics*. Soc. Industr. Appl. Math., Philadelphia.

Bailey, R.A. (1981) A unified approach to design of experiments. *J.R. Statist. Soc.* A **144**, 214–223.

Bailey, R.A. and Rowley, C.A. (1987) Valid randomization. *Proc. Roy. Soc.* (*London*) A **410**, 105–124.

Baker, G.A. and Graves-Morris, P. (1981) *Padé Approximants: I, Basic Theory*. Encycl. of Mathematics and Applications. Addison-Wesley, Reading, Mass.

Barndorff-Nielsen, O.E. (1978) *Information and Exponential Families in Statistical Theory*. Wiley, Chichester.

Barndorff-Nielsen, O.E. (1983) On a formula for the distribution of the maximum likelihood estimator. *Biometrika*, **70**, 343–365.

Barndorff-Nielsen, O.E. (1988) *Parametric Statistical Models and Likelihood.* Lecture notes in statistics. Springer, Heidelberg.

Barndorff-Nielsen, O.E. and Blæsild, P. (1983) Exponential models with affine dual foliations. *Ann. Statist.*, **11**, 753–769.

Barndorff-Nielsen, O.E. and Blæsild, P. (1988a) Combination of reproductive models. *Ann. Statist.*, **16**, 323–341.

Barndorff-Nielsen, O.E. and Blæsild, P. (1988b) Inversion and index notation. Research Report 172, Dept. Theor. Statist., Aarhus University.

Barndorff-Nielsen, O.E. and Cox, D.R. (1979) Edgeworth and saddle-point approximations with statistical applications (with discussion). *J.R. Statist. Soc.* B **41**, 279–312.

Barndorff-Nielsen, O.E. and Hall, P. (1988) On the level-error after Bartlett adjustment of the likelihood ratio statistic. *Biometrika*, **75**, 374–388.

Bartlett, M.S. (1935) Effect of non-normality on the t-distribution. *Proc. Camb. Phil. Soc.*, **31**, 223–231.

Bartlett, M.S. (1938) The characteristic function of a conditional statistic. *J. London Math. Soc.*, **13**, 62–67.

Beard, R.E., Pentikäinen, T. and Pesonen, E. (1984) *Risk Theory.* 3rd edn. Chapman and Hall, London.

Benson, F. (1949) A note on the estimation of mean and standard deviation from quantiles. *J.R. Statist. Soc.* B **11**, 91–100.

Bhattacharya, R.N. (1985) Some recent results on Cramér-Edgeworth expansions with applications. In P.R. Krishnaiah (ed.), *Multivariate Analysis – VI.* Elsevier, Amsterdam, pp. 57–75.

Bhattacharya, R.N. and Ghosh, J.K. (1978) On the validity of the formal Edgeworth expansion. *Ann. Statist.*, **6**, 434–451.

Bhattacharya, R.N. and Rao, R.R. (1976) *Normal approximation and asymptotic expansions.* Wiley, New York. Reprint with corrections and supplemental material (1986). Krieger, Malabar, Florida.

Bishop, Y.M.M., Fienberg, S.E. and Holland, P.W. (1975) *Discrete Multivariate Analysis.* MIT Press, Cambridge, Mass.

Blæsild, P. and Jensen, J.L. (1985) Saddlepoint formulas for reproductive exponential models. *Scand. J. Statist.*, **12**, 193–202.

Bleistein, N. and Handelsman, R.A. (1975) *Asymptotic Expansions of Integrals.* Holt, Rinehart and Winston, New York.

Block, H.W. and Fang, Z. (1988) A multivariate extension of Hoeffding's lemma. To appear in *Ann. Probab.*

Bolthausen, E. (1986) Laplace approximations for sums of independent random vectors. *Probab. Th. Rel. Fields*, **72**, 305–318.

Bolthausen, E. (1987) Laplace approximations for sums of independent random vectors. Part II. Degenerate maxima and manifolds of maxima. *Probab. Th. Rel. Fields*, **76**, 167–206.

Brillinger, D.R. (1969) The calculation of cumulants via conditioning. *Ann. Inst. Statist. Math.*, **21**, 375–390.

Cameron, D. (1986) Euler and MacLaurin made easy. *Math. Sci.*, **12**, 3–20.

Chaganty, N.R. and Sethuraman, J. (1986) Multidimensional large deviation local limit theorems. *J. Mult. Anal.*, **20**, 190–204.

Chandra, T.P. (1985) Asymptotic expansions of perturbed chi-square variables. *Sankhyā* A **47**, 100–110.

Chandra, T.P. and Ghosh, J.K. (1979) Valid asymptotic expansions for the likelihood ratio and other perturbed chi-square variables. *Sankhyā* A **41**, 22–47.

Chihara, T.S. (1978) *An Introduction to Orthogonal Polynomials.* Gordon and Breach, New York.

Cornish, E.A. and Fisher, R.A. (1937) Moments and cumulants in the specification of distributions. *Int. Statist. Review*, **5**, 307–322.

Cox, D.R. (1970) The continuity correction. *Biometrika*, **57**, 217–218.

Cox, D.R. (1984) Long-range dependence: a review. In H.A. David and H.T. David (eds), *Statistics: an appraisal.* Iowa State University Press, Ames, Iowa, pp. 55–74.

Cox, D.R. and Isham, V. (1978) Series expansions for the properties of a birth process of controlled variability. *J. Appl. Prob.*, **15**, 610–616.

Cox, D.R. and Isham, V. (1980) *Point Processes.* Chapman and Hall, London.

Cox, D.R. and Reid, N. (1987) Approximations to noncentral distributions. *Can. J. Statist.*, **15**, 105–114.

Cramér, H. (1937) *Random Variables and Probability Distributions.* Cambridge University Press.

Cramér, H. (1963) On asymptotic expansions for sums of independent random variables with a limiting stable distribution. *Sankhyā* A **25**, 13–24.

Cramér, H. and Leadbetter, M.R. (1967) *Stationary and Related Stochastic Processes.* Wiley, New York.

Daniels, H.E. (1954) Saddlepoint approximations in statistics. *Ann. Math. Statist.*, **25**, 631–650.

Daniels, H.E. (1958) Discussion of paper by D.R. Cox. *J.R. Statist. Soc.* B **20**, 236–238.

Daniels, H.E. (1960) Approximate solutions of Green's type for univariate stochastic processes. *J.R. Statist. Soc.* B **22**, 376–401.

Daniels, H.E. (1982) The saddlepoint approximation for a general birth process. *J. Appl. Prob.*, **19**, 20–28.

Daniels, H.E. (1983) Saddlepoint approximations for estimating equations. *Biometrika*, **70**, 89–96.

Daniels, H.E. (1987) Tail probability approximations. *Int. Statist. Review*, **55**, 37–48.

Darroch, J.N., Lauritzen, S.L. and Speed, T.P. (1980) Markov fields and log-linear interaction models for contingency tables. *Ann. Statist.*, **8**, 522–539.

Davis, A.W. (1976) Statistical distributions in univariate and multivariate Edgeworth populations. *Biometrika*, **63**, 661–670.

de Bruijn, N.G. (1958) *Asymptotic Methods in Analysis.* North-Holland, Amsterdam.

de Haan, L. and Resnick, S.I. (1977) Limit theory for multivariate sample extremes. *Z. Wahrscheinlichkeitstheorie verw. Gebiete*, **40**, 317–337.

Dieudonné, J. (1968) *Calcul Infinitésimal.* Hermann, Paris.

Does, R.J.M.M. (1983) An Edgeworth expansion for simple linear rank statistics under the null-hypothesis. *Ann. Statist.*, **11**, 607–624.

Durbin, J. (1980) Approximations for densities of sufficient statistics. *Biometrika*, **67**, 311–333.

Dyson, F.J. (1943) A note on kurtosis. *J.R. Statist. Soc.*, **106**, 360–361.

Eckhaus, W. (1973) *Matched Asymptotic Expansions and Singular Perturbations.* North-Holland, Amsterdam.

Ellis, R.S. (1985) *Entropy, Large Deviations, and Statistical Mechanics.* Springer, New York.

Erdélyi, A., Magnus, W., Oberhettinger, F. and Tricomi, F.G. (1953) *Higher Transcendental Functions*, Vol. 2. McGraw-Hill, New York.

Esscher, F. (1932) On the probability function in the collective theory of risk. *Skand. Akt. Tidsskr.*, **15**, 175–195.

Esscher, F. (1963) On approximate computation of distribution functions when the corresponding characteristic functions are known. *Skand. Akt. Tidsskr.*, 78–86.

Esséen, C.G. (1945) Fourier analysis of distribution functions. *Acta Math.*, **77**, 1–125.

Feller, W. (1966) *An Introduction to Probability Theory and Its Applications*, Vol. 2. Wiley, New York.

Fernholz, L.T. (1988) Statistical functionals. In S. Kotz and N.L. Johnson (eds) *Encycl. Statist. Sci.*, **8**. Wiley, New York, pp. 656–660.

Field, C.A. and Hampel, F.R. (1982) Small-sample asymptotic distributions of M-estimates of location. *Biometrika*, **69**, 29–46.

Fisher, R.A. (1921) On the 'probable error' of a coefficient of correlation from a small sample. *Metron*, **1**, 3–32.

Fisher, R.A. (1929) Moments and product moments of sampling distributions. *Proc. London Math. Soc.* Series 2, **30**, 199–238.

Fisher, R.A. (1934) Two new properties of mathematical likelihood. *Proc. Roy. Soc. (London)* A **144**, 285–307.

Föllmer, H. (1987) Random fields and diffusion processes. In P.L. Hennequin (ed.), *École d'Été de Probabilités de Saint Flour XVI*. Lecture Notes in Mathematics. Springer, Berlin.

Freidlin, M.I. and Wentzell, A.D. (1984) *Random Perturbations of Dynamical Systems.* Springer, New York.

Galambos, J. (1978) *The Asymptotic Theory of Extreme Order Statistics.* Wiley, New York.

Gill, R.D. (1987) Non- and semiparametric maximum likelihood estimators and the von Mises method (part I). Research Report MS-R8709, Centre Math. Comp. Sci., Amsterdam.

Gnedenko, B.V. (1962) *The Theory of Probability.* Chelsea, New York.

Götze, F. and Hipp, C. (1983) Asymptotic expansions for sums of weakly dependent random variables. *Z. Wahrscheinlichkeitstheorie verw. Gebiete*, **64**, 211–239.

Grab, E.L. and Savage, I.R. (1954) Tables of the expected value of 1/X for positive Bernoulli and Poisson variables. *J. Amer. Statist. Assoc.*, **49**, 169–177.

Grad, H. (1949) Note on the N-dimensional Hermite polynomials. *Comm. Pure Appl. Math.*, **3**, 325–330.

Groeneboom, P. and Oosterhoff, J. (1981) Bahadur efficiency and small-sample efficiency. *Int. Statist. Review*, **49**, 127–141.

Grundy, P.M. and Healy, M.J.R. (1950) Restricted randomization and quasi-Latin squares. *J.R. Statist. Soc.* B **12**, 286–291.

Guillemin, V. and Sternberg, S. (1977) *Geometric Asymptotics.* Mathematical Surveys, No. 14. American Mathematical Society, Providence, Rhode Island.

Hájek, J. (1968) Asymptotic normality of simple linear rank statistics under alternatives. *Ann. Math. Statist.*, **39**, 325–346.

Hald, A. (1981) T.N. Thiele's contributions to statistics. *Int. Statist. Review*, **49**, 1–20.

Hall, M. (1986) *Combinatorial Theory.* (2nd edn). Wiley, New York.

Hida, T. (1980) *Brownian Motion.* Springer, Berlin.

Hipp, C. (1984) Asymptotic expansions for conditional distributions: the lattice case. *Probab. Math. Statist.*, **4**, 207–219.

Hoeffding, W. (1948) A class of statistics with asymptotically normal distributions. *Ann. Math. Statist.*, **19**, 293–325.

Höglund, T. (1970) On the convergence of convolutions of distributions with regularly varying tails. *Z. Wahrscheinlichkeitstheorie verw. Gebiete*, **15**, 263–272.

Holst, L. (1979) Two conditional limit theorems with applications. *Ann. Statist.*, **7**, 551–557.

Holst, L. (1981) Some conditional limit theorems in exponential families. *Ann. Probab.*, **9**, 818–830.

Hougaard, P. (1982) Parameterizations of non-linear models. *J.R. Statist. Soc.* B **44**, 244–252.

Hsu, L.C. (1948) A theorem on the asymptotic behaviour of a multiple integral. *Duke Math. J.*, **15**, 623–632.

Irwin, J.O. and Kendall, M.G. (1944) Sampling moments of moments for a finite population. *Ann. Eugen.*, **12**, 138–142.

Isham, V. (1981) An introduction to spatial point processes and Markov random fields. *Int. Statist. Review*, **49**, 21–43.

Isobe, E. and Sato, S. (1983) Wiener-Hermite expansion of a process generated by an Itô stochastic differential equation. *J. Appl. Prob.*, **20**, 754–765.

Itô, K. (1951) Multiple Wiener integral. *J. Math. Soc. Japan*, **3**, 157–169.

James, G.S. (1958) On moments and cumulants of systems of statistics. *Sankhyā*, **20**, 1–30.

James, G.S. and Mayne, A.J. (1962) Cumulants of functions of random variables. *Sankhyā*, **24**, 47–54.

Janson, S. (1988) Normal convergence by higher semiinvariants with applications to sums of dependent random variables and random graphs. *Ann. Probab.*, **16**, 305–312.

Jensen, J.L. (1981) On the hyperboloid distribution. *Scand. J. Statist.*, **8**, 193–206.

Jensen, J.L. (1986) Validity of the formal Edgeworth expansion when the underlying distribution is partly discrete. Research Report 148, Dept. Theor. Statist., Institute of Mathematics, Aarhus University.

Jensen, J.L. (1987) Standardized log-likelihood ratio statistics for mixtures of discrete and continuous observations. *Ann. Statist.*, **15**, 314–324.

Jensen, J.L. (1988a) On asymptotic expansions in non-ergodic models. *Scand. J. Statist.*, **14**, 305–318.

Jensen, J.L. (1988b) Uniform saddlepoint approximations. *Adv. Appl. Prob.*, **20**, 622–634.

Jørgensen, B. (1982) *Statistical Properties of the Generalized Inverse Gaussian Distribution.* Lecture Notes in Statistics, 9. Springer, New York.

Kampé de Fériet, J. (1923) Sur une formule d'addition des polynomes d'Hermite. *Det Kgl. Danske Vidensk. Sel. Mat.-fys. Medd.*, **5**, 2.

Kendall, D.G. (1942) A summation formula associated with finite trigonometric integrals. *Q.J. Math.*, **13**, 172–184.

Kendall, M.G. and Stuart, A. (1969) *Advanced Theory of Statistics*, Vol. 1. (3rd edn). Griffin, High Wycombe.

Khinchin, A.I. (1929) Über einen neuen Grenzwertsatz der Wahrscheinlichkeitsrechnung. *Math. Annalen*, **101**, 745–752.

Kiiveri, H., Speed, T.P. and Carlin, J.B. (1984) Recursive causal models. *J. Austral. Math. Soc.* A **36**, 30–52.

Kudlaev, E.M. (1984) Limiting conditional distributions for sums of random variables. *Th. Probab. Applic.*, **29**, 776–786.

Lancaster, H.O. (1969) *The Chi-squared Distribution.* Wiley, New York.

Landers, D. and Rogge, L. (1986) Second order approximation in the conditional central limit theorem. *Ann. Probab.*, **14**, 313–325.

Landers, D. and Rogge, L. (1987) Nonuniform estimates in the conditional central limit theorem. *Ann. Probab.*, **15**, 776–782.

Lauritzen, S.L. (1981) Time series analysis in 1880: A discussion of contributions made by T.N. Thiele. *Int. Statist. Review*, **49**, 319–331.

Lauritzen, S.L. (1982) *Lectures on Contingency Tables.* Aalborg University Press.

Lauritzen, S.L. and Wermuth, N. (1989) Graphical models for associations between variables, some of which are qualitative and some quantitative. To appear in *Ann. Statist.*

Leadbetter, M.R., Lindgren, G. and Rootzén, H. (1983) *Extremes and Related Properties of Random Sequences and Processes.* Springer, New York.

Leonov, V.P. and Shiryaev, A.N. (1959) On a method of calculation of semi-invariants. *Theor. Probab. Applic.*, **4**, 319–329.

Lindley, D.V. (1961) The use of prior probability distributions in statistical inference and decisions. *Proc. 4th Berkeley Symp.*, **1**, 453–468.

Malyshev, V.A. (1980) Cluster expansions in lattice models of statistical physics and the quantum theory of fields. *Russian Math. Surveys*, **35** (2), 1–62.

Mann, H.B. and Wald, A. (1943) On stochastic limit and order relationships. *Ann. Math. Statist.*, **14**, 217–226.

Martin-Löf, A. (1986) Entropy, a useful concept in risk theory. *Scand. Actuarial J.*, 223–235.

McCullagh, P. (1984) Tensor notation and cumulants of polynomials. *Biometrika*, **71**, 461–476.

McCullagh, P. (1987) *Tensor Methods in Statistics*. Chapman and Hall, London.

McCullagh, P. and Wilks, A.R. (1988) Complementary partitions. *Proc. R. Soc. (London)* A **415**, 347–362.

Michel, R. (1979) Asymptotic expansions for conditional distributions. *J. Multivariate Anal.*, **9**, 393–400.

Molenaar, W. (1970) Approximation to the Poisson, Binomial and Hypergeometric Distribution Functions. *Math. Centrum Tracts*, **31**. Amsterdam.

Moolgavkar, S.H. and Venzon, D.J. (1987) Confidence regions in curved exponential families: application to matched case-control and survival studies with general relative risk function. *Ann. Statist.*, **15**, 346–359.

Moran, P.A.P. (1950) Numerical integration by systematic sampling. *Proc. Camb. Phil. Soc.*, **46**, 111–115.

Moran, P.A.P. (1968) *An Introduction to Probability Theory*. Clarendon Press, Oxford.

Niki, N. and Konishi, S. (1986) Effects of transformations in higher order asymptotic expansions. *Ann. Inst. Statist. Math.*, **38**, 371–383.

Ogata, Y. and Tanemura, M. (1984) Likelihood analysis of spatial point patterns. *J.R. Statist. Soc.* B **46**, 496–518.

Olver, F.J. (1974) *Asymptotics and Special Functions*. Academic Press, New York.

Pearson, E.S. and Hartley, H.O. (1966) *Biometrika Tables for Statisticians*, Vol. 1 (3rd edn). Biometrika Trust.

Pedersen, B.V. (1979) Approximating conditional distributions by the mixed Edgeworth-saddlepoint expansion. *Biometrika*, **66**, 597–604.

Phillips, P.J.B. (1982) Best uniform and modified Padé approximants to probability densities in econometrics. In W. Hilderbrand (ed.), *Advances in Econometrics*, Cambridge University Press, pp. 123–167.

Plackett, R.L. (1954) A reduction formula for normal multivariate integrals. *Biometrika*, **41**, 351–360.

Reeds, J.A. (1976) On the definition of von Mises functionals. Research Report S-44, Dept. Statistics, Harvard University.

Reid, N. (1983) Influence functions. In S. Kotz and N.L. Johnson (eds) *Encycl. Statist. Sci.*, **4**, Wiley, New York, pp. 117–119.

Reid, N. (1988) Saddle-point expansions and statistical inference (with discussion). To appear in *Statistical Science*, **3**.

Renyi, A. (1970) *Probability Theory.* North-Holland, Amsterdam.

Rice, J.R. (1964) *The Approximation of Functions*, Vol. 1. Addison Wesley, New York.

Rice, J.R. (1969) *The Approximation of Functions*, Vol. 2. Addison Wesley, New York.

Robinson, J. (1982) Saddlepoint approximations for permutation tests and confidence intervals. *J.R. Statist. Soc.* B **44**, 91–101.

Rockafellar, R.T. (1970) *Convex Analysis.* Princeton University Press.

Rota, G.-C. (1964) On the foundations of combinatorial theory I. Theory of Möbius functions. *Z. Wahrscheinlichkeitstheorie verw. Gebiete*, **2**, 340–368.

Ruelle, D. (1969) *Statistical Mechanics.* Benjamin, Reading, Mass.

Schilder, M. (1966) Some asymptotic formulae for Wiener integrals. *Trans. Amer. Math. Soc.*, **125**, 63–85.

Serfling, R.J. (1980) *Approximation Theorems of Mathematical Statistics.* Wiley, New York.

Shohat, J.A. and Tamarkin, J.D. (1943) *The Problem of Moments.* American Mathematical Society, Providence, Rhode Island.

Simmonds, J.G. and Mann, J.E. (1986) *A First Look at Perturbation Theory.* Krieger, Malabar, Florida.

Skovgaard, I.M. (1981) Transformation of an Edgeworth expansion by a sequence of smooth functions. *Scand. J. Statist.*, **8**, 207–217.

Skovgaard, I.M. (1986) On multivariate Edgeworth expansions. *Int. Statist. Review*, **54**, 169–186.

Skovgaard, I.M. (1987) Saddlepoint expansions for conditional distributions. *J. Appl. Prob.*, **24**, 875–887.

Slepian, D. (1962) The one-sided barrier problem for Gaussian noise. *Bell System Tech. J.*, **41**, 463–501.

Speed, T.P. (1983) Cumulants and partition lattices. *Austral. J. Statist.*, **25**, 378–388.

Speed, T.P. (1986a) Cumulants and partition lattices II. Generalised *k*-statistics. *J. Austral. Math. Soc.* A **40**, 34–53.

Speed, T.P. (1986b) Cumulants and partition lattices III. Multiply-indexed arrays. *J. Austral. Math. Soc.* A **40**, 161–182.

Speed, T.P. (1986c) Cumulants and partition lattices IV. A.S. convergence of generalised k-statistics. *J. Austral. Math. Soc.* A **41**, 79–94.

Speed, T.P. (1987) What is an analysis of variance? (With discussion.) *Ann. Statist.*, **15**, 885–941.

Speed, T.P. and Silcock, H.L. (1985a) Cumulants and partition lattices V. Calculating generalised k-statistics. *J. Austral. Math. Soc.* A **44**, 171–196.

Speed, T.P. and Silcock, H.L. (1985b) Cumulants and partition lattices VI. Variances and co-variances of mean squares. *J. Austral. Math. Soc.* A **44**, 362–388.

Stanley, R.P. (1986) *Enumerative Combinatorics.* Vol. I. Wadsworth and Brooks/Cole, Pacific Grove, California.

Steck, G.P. (1957) Limit theorems for conditional distributions. *Univ. California Publ. Statist.*, **2**, 237–284.

Stroock, D.W. (1984) *An Introduction to the Theory of Large Deviations.* Springer, New York.

Swensen, A.R. (1983) A note on convergence of distributions of conditional moments. *Scand. J. Statist.*, **10**, 41–44.

Szegö, G. (1967) *Orthogonal Polynomials* (3rd edn). American Mathematical Society, Providence, Rhode Island.

Taqqu, M.S. (1974) Weak convergence to fractional Brownian motion and to the Rosenblatt process. *Z. Wahrscheinlichkeitstheorie verw. Gebiete*, **31**, 287–302.

Taqqu, M.S. (1979) Convergence of integrated processes of arbitrary Hermite rank. *Z. Wahrscheinlichkeitstheorie verw. Gebiete*, **50**, 53–83.

Temme, N.M. (1982) The uniform asymptotic expansion of a class of integrals related to cumulative distribution functions. *SIAM J. Math. Anal.*, **13**, 239–253.

Temme, N.M. (1983) Uniform asymptotic expansions of Laplace integrals. *Analysis*, **3**, 221–249.

Temme, N.M. (1985) Special functions as approximants in uniform asymptotic expansions of integrals. In *Special Functions: Theory and Computation.* Rendiconti Seminario Matematico, Torino, pp. 289–317.

Thiele, T.N. (1889) *Forelaesninger over Almindelig Iagttagelseslære: Sandsynlighedregning og mindste Kvadraters Methode.* Reitzel, København.

Thiele, T.N. (1897) *Elementaer Iagttagelseslære.* Gyldendalske, København. English translation: *Theory of Observations.* (1903): Layton, London. Reprinted in *Ann. Math. Statist.* (1931) **2**, 165–308.

Thiele, T.N. (1899) Om Iagttagelseslærens Halvinvarianter. *Overs. Vid. Sels. Forh. Nr.*, **3**, 135–141.

Tiago de Oliveira, J. (1984) Bivariate models for extremes: statistical decision. In J. Tiago de Oliveira (ed.), *Statistical Extremes and Applications.* Reidel, Dordrecht, pp. 131–153.

Tierney, L. and Kadane, J.B. (1986) Accurate approximations for posterior moments and marginal densities. *J. Amer. Statist. Assoc.*, **81**, 82–86.

Tiku, M.L. (1964) A note on the negative moments of a truncated Poisson variable. *J. Amer. Statist. Assoc.*, **59**, 1220–1224.

Tjur, T. (1984) Analysis of variance in orthogonal designs. *Int. Statist. Review*, **52**, 33–65.

Tukey, J.W. (1950) Some sampling simplified. *J. Amer. Statist. Assoc.*, **45**, 501–519.

van den Berg, I. (1987) *Nonstandard Asymptotic Analysis.* Lecture Notes in Mathematics. Springer, Berlin.

van Dyke, M. (1975) *Perturbation Methods in Fluid Mechanics.* Parabolic Press, Stanford, California.

van Zwet, W.R. (1984) A Berry-Esseen bound for symmetric statistics. *Z. Wahrscheinlichkeitstheorie verw. Gebiete*, **66**, 425–440.

Varadhan, S.R.S. (1984) *Large Deviations and Applications.* CBMS-NSF Regional Conference Series in Applied Mathematics. SIAM, Philadelphia.

Volterra, V. (1931) *Theory of Functionals.* Blackie, London.

Weiss, G.H. and Kiefer, J.E. (1983) The Pearson random walk with unequal step sizes. *J. Phys.* A **16**, 489–495.

Withers, C.S. (1983) Expansions for the distribution and quantiles of a regular functional of the empirical distribution with applications to nonparametric confidence intervals. *Ann. Statist.*, **11**, 577–587.

Withers, C.S. (1984) Asymptotic expansions for distributions and quantiles with power series cumulants. *J.R. Statist. Soc.* B **46**, 389–396.

Yates, F. (1948) Systematic sampling. *Phil. Trans. R. Soc.* (*London*) A **241**, 345–377.

Yates, F. (1951) Bases logiques de la planification des expériences. *Ann. Inst. H. Poincaré*, **12**, 97–112.

Author index

Subject index